W9-DJK-019

In the series

Animals, Culture, and Society

edited by

Clinton R. Sanders and Arnold Arluke

Fishy Business

Salmon, Biology, and the Social Construction of Nature

Fishy Business

Salmon, Biology, and the Social Construction of Nature

RIK SCARCE

TEMPLE UNIVERSITY PRESS
Philadelphia

Temple University Press, Philadelphia 19122

Published 2000
Printed in the United States of America

⊗ The paper used in this publication meets the requirements of the American National Standard for Information Sciences—Permanence of Paper for Printed Library Materials, ANSI Z39.48–1984

Library of Congress Cataloging-in-Publication Data

Scarce, Rik, 1958–
Fishy business : salmon, biology, and the social construction of nature / Rik Scarce.
p. cm. — (Animals, culture, and society)
Includes bibliographical references and index.
ISBN 1-56639-728-6 (cloth : alk. paper). — ISBN 1-56639-729-4 (paper : alk. paper)
1. Nature—Effect of human beings on—United States. 2. Human ecology—United States—Philosophy. 3. Salmon—United States—Physiology. 4. Salmon—Ecology—United States. 5. Salmon fisheries—United States. 6. United States—Environmental conditions. I. Title. II. Series.
GF503.S23 1999
304.2—dc21 99–24029

For my wife, Petra,

my son, Alexander,

and for all those who helped make this possible,
each in his or her own way:

Loren,
Lew,
Noël,
Armand,
and my mother

Contents

Acknowledgments

My thanks to the following for permission to reprint portions of my previous work:

From "What Do Wolves Mean? Conflicting Constructions of *Canis lupus* in 'Bordertown,'" *Human Dimensions of Wildlife* 3, no. 3 (1998): 26–45, courtesy of the journal.

From "Socially Constructing Pacific Salmon," *Society and Animals* 5, no. 2 (1997): 117–35, courtesy of the journal.

And from "Who—or What—Is in Control Here? The Social Context of Salmon Biology," *Society and Natural Resources* 12, no. 8 (1999): 763–76, courtesy of Taylor and Francis Publishers.

1 Nature in the Making

IN THE SUMMER OF 1993, the movie *Jurassic Park* took hold of American culture. Even those who dislike pulp fiction or pop movies could not miss the endless monologues on late night television, lengthy magazine articles, and cocktail party discourses about dinosaurs brought back to life by ingenious and thrilling—or were they despicable and criminal?—manipulations of DNA extracted from the bodies of prehistoric mosquitoes trapped in amber. Aside from the film version's unsuitability for younger viewers, the biggest controversy surrounding *Jurassic Park* was the scientific validity of its premise: namely, that intact dinosaur DNA 100 million years old could indeed be found and used to clone dinosaurs—bring them back to life.

Yet that was only the scientific premise of *Jurassic Park.* Author and screenwriter Michael Crichton had a more provocative *social* ax to grind. Late in the book version of the story, Crichton's skeptical mathematician said, "Science has attained so much power that its practical limits begin to be apparent. Largely through science, billions of us live in one small world, densely packed and intercommunicating. But science cannot help us decide what to do with that world, or how to live. . . . Ever since Newton and Descartes, science has explicitly offered us the vision of total control. Science has claimed the power to eventually control everything, through its understanding of natural laws. But in the twentieth century, that claim has been shattered beyond repair."[1] Entertaining though *Jurassic Park* was, in these words Crichton challenged us to consider how we humans exert control over this thing we call "nature," specifically through our clumsy use of science and technology. For all of science's ostensible objectivity and exactness, it can yield terrifying results when control is lost—and even when it is successfully applied and maintained. Most important, science cannot aid society in establishing norms—socially expected and accepted attitudes and behaviors—or in making value-based decisions about nature. Science rests upon a foundation that identifies it alone as the privileged instrument for knowledge creation about nature, but there is no allowance in scientific objectivity for evaluating the ethical questions that arise out of science.

Moreover, *Jurassic Park* calls into question what nature is in the first place. Are cloned beings natural? Is it natural to bring plants and animals extinct for millions of years into the present? That is the departure point of this book: What, indeed, is *nature?* Such a question is actually one of meaning and even reality. How is nature defined socially, and how has it come to exist at all? Is nature the same to everyone? And, if it is not eternal or consistent in its meaning, whose nature is the dominant one and what are the processes by which nature's meaning changes? Even before we can ask the ethical and normative questions about nature posed in *Jurassic Park,* we need to come to terms with what nature is in a social sense by asking questions like these.

NATURE'S BEGINNINGS

Ultimately, these questions compel us to answer which nature it is that society actually is creating and which one it should be creating. This search for the social meaning—*meanings,* really—of nature is far more complex than looking up the word in a dictionary, as we shall see. To begin this discussion, I offer the following observation: nature is four thousand years old. Historians and etymologists tell us that nature as an identifiable concept has existed in the Western tradition since the Greeks created the term *phuysis,* which initially meant "everything." This meaning was "passed to *natura,*" the Latin root of our word "nature," and nature's denotation further evolved through the centuries.[2]

As a concept, then, nature has not always been there. Nor is it found in all cultures. Some students and I once were discussing the nature concept. We had gotten about as far along in the story as I have here—the seed of skepticism regarding nature had been planted and brows were deeply furrowed in a combination of curiosity and confusion—when one student spoke what the others must have been thinking. "I don't believe you," she said. "Of course every culture has a concept like nature. Nature obviously is everywhere around us and around every other society." Later that day I asked colleagues who speak Crow and Cheyenne about "nature" in their languages and cultures. Both said there is no term that translates even roughly into our "nature." Nor are those cultures alone, as geographer Ian Simmons has noted.[3] Nature is not always and everywhere present, but exists only when societies conceptually distinguish themselves from their surroundings.

At some level nature must be a social creation. Today, *Nature*—this connotative, meaning-filled, and socially created nature that I have in mind when I speak of it as "socially constructed"—is constantly being remade by us, sometimes through intense conflict, such as seen in efforts to "save the whales," "save the rain forests," and even "save native peoples," and sometimes in more subtle ways, such as alterations in laws allowing or prohibiting pollution and timber cutting. Even the North American Free Trade Agreement was the topic of intense debate over what it would mean for "the environment," a synonym for Nature that I will also capitalize to highlight its contingent character. In all of these debates, science has a prominent place in the changing meanings of Nature and the Environment.[4] Politicians, corporations, and even the most radical, antitechnological environmental groups invoke science to support their unique conceptualizations of Nature.[5]

SCIENTISTS, RIVERS, AND SALMON

This study stands at the crossroads of Nature and society's primary tool for investigating Nature, namely, science. It explores how one part of Nature, Pacific salmon, is socially constructed by one group of scientists, salmon biologists. Social constructions are the meanings of things that result from interactive social processes, and our interest here is the interactions between salmon biologists, various social institutions, and the salmon. Salmon biologists pursue their work in many ways, and they do so in a host of venues. Some study "life history" processes, such as juvenile salmon incubation, fresh and salt water migration, the survival of fish in different locations, and salmons' relationships to predators. Others emphasize genetics, and still others "stock recruitment"—increasing the numbers of fish available to be caught by humans. A small fraction of salmon biologists are consultants or work for corporations; most are employed by governments or Indian tribes, or they are affiliated with universities. Many of these biologists focus their attention on salmon living in rivers and the Pacific Ocean, although some spend virtually all of their careers in laboratories or in front of computer screens creating mathematical models from the data that their colleagues have gathered.

The rivers of most concern here are the Columbia and the Snake in the United States, although on occasion I will mention the Fraser River in southern British Columbia as well. Limnologists see the Columbia and the Snake as a single river system. The Snake rises in Wyoming and drains portions of Montana, Idaho, Utah, Nevada, and Oregon before it reaches the Columbia in Washington State. In his book *The Snake River*, river advocate Tim Palmer neatly presents the vital statistics: "The Snake is the tenth longest river in the United States. Carrying 37 million acre feet a year, it exceeds by two and a half times the volume of the Colorado River (1 acre foot covers an acre with 1 foot of water). The Snake receives 30 percent of the runoff from the eight mountain states and drains much of the northwestern Rockies, or 109,000 square miles, an area larger than Colorado. The basin is roughly 450 miles in length and width."[6] The Snake's canyons are spectacular, its roaring waterfalls are magnificent, and its salmon are all but gone, thanks to dams, irrigation, habitat destruction of many kinds, the impacts of hatchery-raised and non-native fish, and overfishing.

The Columbia shares many similarities with the Snake. In addition to their stunning scenery and the human-caused blows to the rivers and the salmon living there, both flow through high desert, almost all of which receives less than two feet of precipitation annually. For the most part human population is sparse; when compared to the Hudson, Connecticut, or Mississippi drainages, the human presence in terms of raw numbers borders on the insignificant, though the human impacts are as great as those on any eastern river. Near its lower reaches the Columbia rolls past Portland, the only large city along its entire 1,270-mile length, though it or its tributaries pass through Oregon's lush Willamette Valley and Washington cities including Richland, Kennewick, Pasco, Yakima, and Spokane.

Anthony Netboy began a book on the river's salmon by writing, "The Columbia River rises in Columbia Lake, some 80 miles north of the United States border in British Columbia at an elevation of 2,650 feet above sea level. . . . [It] enters the United States below Trail, in the northeast corner of the state of Washington, thus completing a journey of about 460 miles and dropping 1,360 feet in elevation."[7] The Snake is so large that when combined with the Columbia it makes a monstrously large river draining 259,000 square miles, and its mouth at the Pacific Ocean is five miles wide.

The Columbia-Snake and the salmon inhabiting it evolved together, and there were perhaps 16 million adult salmon spawning in the river and its tributaries annually when Lewis and Clark visited in the first decade of the nineteenth century. The seven species of Pacific salmon found in North America are known by evocative names like sockeye, king, chum, silver, pink, steelhead, and cutthroat; two other species live only in Asia. The genus name for all nine is *Oncorhynchus,* Latin for "hook-nosed" from the curvature that male salmons' jaws take on at spawning time.

In a thumbnail sketch, this is the life course of salmon "in the wild" as scientists understand it: Born in gravelly streams carved by glaciers, after emerging from their stony nursery the fry may spend virtually no time at all in their natal waters or they may remain there for up to two years. Eventually, they swim toward the ocean as "smolts," undergoing extensive transformations that alter their body chemistry and enable them to live in salt water. After one to five years of maturing in the ocean, where their weight increases a hundredfold, their body chemistry again changes, this time reversing itself so that the salmon are adapted to living in fresh water. They swim inland to the same stream where they were born, in some instances traveling nine hundred miles to reach spawning grounds seven thousand feet above sea level. There they mate and, with the exception of some steelhead and cutthroat trout, the adults die shortly thereafter (the trout were only recently reclassified as salmon; even among biologists the trout label persists).

Historically, human-salmon interactions have been crucial to every society living within the salmons' range. For example, salmon played a central role in pre-Columbian cultures. Rituals developed around the fish, and they were worshiped as gods. And for good reason: economically the salmon were a product for trade, and these key religious icons provided the source of much of the sustenance for numerous tribes. Moreover, for many of these societies Nature as such did not exist. The absence of such a word in numerous native North American languages, like the Crow and Cheyenne mentioned earlier, reflects the fact that those cultures did not distinguish between themselves and other beings. Salmon and people were one.

The Europeans who successfully pursued the conquest of the land and the aboriginal inhabitants of northwest North America brought with them ways of thinking, speaking, and behaving that treated the salmon not as equal actor but as *resource.* The dominant attitude was that the fish were exclusively economic goods available for sale, trade, and manipulation. Offshore fishing, riverwide nets, fish wheels, dams, pollution, and other technologies and technological byproducts simultaneously created and reflected a new form of interaction with salmon. Corresponding with the rise in industrial production was a sharp reduction of the number of salmon in the region. Today, fewer than 10 percent of adult salmon that return to the Columbia and Snake river system to spawn are progeny of wild fish. Most are born and reared in fish hatcheries, highly industrialized production facilities created to grow fish quickly and economically. Wild salmon are so rare that there are dozens of "runs"—or distinct populations—of salmon listed as endangered or threatened under the U.S. Endangered Species Act, and many more are in such trouble that they have been proposed for listing. Canada is rapidly catching up with the United States as its once thriving salmon runs suffer under that nation's lax environmental laws.

WHY SALMON?

Strange as it may sound, I did not select salmon as the subject of this study because they were endangered or because they were prominently in the news. Indeed, I wanted to avoid focusing on salmon as particularly special and merely explore the varying meanings of the fish absent any controversy surrounding them (though when I began this work I had no idea whether biologists would actually attribute more than one meaning to salmon). I suppose earthworms might have been a better choice, since their existence is void of the anxiety we feel about the future of salmon. But there were few earthworm biologists around. Besides, salmon have fascinated me since I was a child; their tragic and monumental journey home from years at sea to spawn and inevitably to die often reminds me of the film *Das Boot,* in which German submariners limp back to port after a harrowing voyage only to be killed by a bombing raid before they set foot on shore.

As my research proceeded, however, I found that there is much more to the salmon-human story than mere pathos. Salmon embody a unique nexus of social value and scientific curiosity. Their cultural importance extends to many societies, both historically and contemporaneously. They are symbols of food, of fun, of a higher power. They are extraordinarily valuable economically and are the stuff that nations go to war over (even if the wars have not been very "hot"). Scientifically, salmon remain a conundrum. We manipulate them to serve our ends, yet they resist, their slippery, wet scales slapping against our best-laid schemes. What other subject of biological study receives such extensive attention from science and society, and is used is so many ways, yet retains so much of what we would like to think is its "wild" character?

CONSTRUCTING NATURE

My concern in the first few chapters of this book is how social psychological, institutional, and technological forces contribute to salmon biologists' constructions of salmon. Social psychology encompasses a range of topics, and the most important of them here are the sorts of interpersonal interactions that we all engage in whenever we come into contact with other persons. Social psychology is especially important to this study because it is through social interaction that old meanings are reinforced and new ones emerge.[8] By "institutions" sociologists mean the formal, nonpersonal entities that shape our lives. Among the institutions and organizations of primary interest here are education (especially universities), government, and the economy. Typically, sociologists emphasize either social psychology or institutions in their studies.[9] This project addresses both of these and *technology* as well. In technologies we find physical, literal embodiments of the figurative, furtive, often unspoken meanings of Nature that pervade society. Technologies enable society to affect Nature and to *effect* it; dams and fish hatcheries are our "prosthetics,"[10] material extensions of society through which we act out and act upon our social constructions of Nature.

Implicit in this tripartite approach is that meanings and knowledge are of little interest unless they are acted upon by individuals, by social groups like scientists, and by institutions such as governments and corporations. Meanings are produced and reproduced by and through social action, including the conversations we have, the organizations we create, and the technologies we use to manipulate the world around us.

That humans construct and give meaning to Nature through their cognitions, institutions, and technologies is an admittedly challenging proposition, especially to extreme "realists" who object to assertions that the nonhuman world is not simply as it appears to us. Sociologist Stephan Fuchs perceptively summarized the realist view, writing, "Ontologically, social and natural worlds are assumed to differ in that the social world, but not nature, is constituted symbolically. Social reality is structured symbolically in that members interpret their own actions and gestures and communicate their interpretations to others. Nature, on the other hand, is sheer physical and observational reality because nature does not talk."[11] Contrast realism with William Cronon's constructivist approach. He has written that environmental historians "understand that the very term they use to describe the environment—*nature*—is itself an astonishingly complex human construction: as Raymond Williams once remarked, 'the idea of nature contains, though often unnoticed, an extraordinary amount of human history.' "[12]

It cannot be emphasized too strongly that *meanings are what are being constructed,* not "things-in-themselves" or pure material reality untouched by human artifice. In other words, when we imbue *anything*—including Nature—with meaning, we alter it. Meanings are dependent on words, and words are only symbols. They are not "the real thing." They are ideas, social realities. When we make a thing mean something, we conceptualize it. We draw it into the human social world through the abstract symbols of language.

The sociological goal of this study is the identification of the social forces that prompt our ideas of salmon and that impose meaning upon them (and upon Nature more generally as well), and the exploration of the impacts of those resulting meanings. The key sociological questions before us include: Do biologists as a group embrace a single meaning of salmon? If not, how do new meanings emerge and how are they negotiated within the discipline? Related to this are questions such as, Which social forces—which institutions and organizations—appear best situated to influence meaning-creating processes? What interests are served by doing so? And how do less powerful social actors effect changes in meaning in the face of dominant, hegemonic forces?

Through the last four millennia, Western societies have constructed a Nature that is fundamentally different and distinct from society. We say through our actions, technologies, and words that our society is not part of Nature. Yet what occurs in the constructing of this distant Nature, and why is it important? This is another question I pose throughout. For now, it is enough to note that those efforts at differentiation create a paradox, as environmental historian Neil Evernden provocatively stated in *The Social Creation of Nature:*

> We exist amidst an accepted dualism: there is Nature, the domain of science, and there is [society], the domain of human action. Nature can be known through scientific explanation, but should human things be "factualized"? To do so is indeed to draw humanity into the domain of Nature and to explain ourselves as the products of "natural law." But to refuse is to reject the acknowledged path to the truth, for Nature is *historically* established as the domain of the knowable, the certain, the absolute. If you want to know the truth, you use the scientific method of investigation. If you want to know the truth about humanity, you make it into an object of scientific investigation as well, a knowable object. But if you do so, you reverse the roles of the two domains, so that Nature, initially a convenient subcategory in the realm of human affairs, becomes the dominant category and absorbs all, including its creator.[13]

What is our connection to Nature? Is Nature just there, available to us to experience immediately and unmediated? Or is it not already the case that we have evidence that Nature is not so universally agreed upon, that it is a formidable task to create and maintain a particular Nature—a task not so easily accomplished? I can foreshadow the answer: Thanks to an array of sociohistorical factors, we disagree sharply over Nature, its place in our lives, and our place in it. But skepticism abounds whenever the social construction of Nature arises, as it should. By emphasizing only one Natural entity (salmon) and only one social group (salmon biologists) I hope to show the tenuousness and contentiousness of our Nature concept in high relief. We shall find that not all salmon are the same, nor are all salmon biologists. The varieties of both have their roots in society and in social processes.

Classical Social Constructivism: An Overview

Before we begin exploring the social construction of Nature via biologists' constructions of salmon, it is important to understand some of the intellectual underpinnings of social constructivism. Constructivism—

or "constructionism," as it is sometimes called—can be thought of as a paradigm, a way of seeing, in that one does not apply constructivism directly to social phenomena as one might apply a theory to data. Constructivism is a general guide to understanding the social world, not a step-by-step formula that claims to predict exactly what will happen. To understand it, several points should be kept in mind.

First, in their now classic book *The Social Construction of Reality,* Peter Berger and Thomas Luckmann took as existential and axiomatic the world around us and did not treat the question of existence—ontology—as itself problematic. They were realists, but not naively so, and their approach contrasts sharply with "idealism," which asserts that "ideas, or thought, are the fundamental reality."[14] In keeping with realism and with the thinking of most sociologists, for Berger and Luckmann there is a world that exists apart from our knowing it. This is a crucial point, and one often overlooked in debates over constructivism, because many neoconstructivists are idealists. Classical constructivists argue that bricks and books do have substance; their existence is real and actual, not a creation of the mind. But their *meaning* is another matter. In a social sense all things are constructed, and we know the world only through our shared meanings of it. It is these socially constructed meanings, and the social processes that give rise to them, that are of interest to constructivists.

Following from this is a second point. Berger and Luckmann's key premise is that social facts are dependent upon the societies out of which they emerge. This leads constructivists to argue that there are no eternal truths. Instead, facts are social products, and they reflect the time and the place in which they are created. So, what is "known" is contextualized by the social situation in which the knowledge is created. Societies, then, *create* knowledge. Knowledge is not just *there* waiting to be discovered or having been discovered. Rather, it is shaped by the society creating the knowledge, leading Berger and Luckmann to write that *knowledge is relative between societies, as it often is even among social groups within the same society.*

An example of the relativity of knowledge, and of the central place of meaning in it, comes from different societies' relationships with salmon. Today we "know" that salmon swim vast distances in the ocean before turning inland to spawn. But to many peoples who inhabited the Pacific Northwest in pre-Columbian times, what we today call "salmon" were *people,* dwellers of the ocean who took on the form of fish and ventured up the rivers.[15] The salmon people were not a "myth" to those tribes. Salmon really were people in fishes' clothing. These salmon people could be born anew and would return in future years to feed the peoples of the land if the bones of the first fish were placed in rivers and allowed to take home word that the land dwellers respected and revered the salmon. The tribes knew this with as much certainty as society today knows the scientific version of the salmon story. The two societies give salmon different respective meanings: biological organisms with very little in common with humans versus people who simply look different when they swim upstream. Social factors such as religion, economics, politics and government, and technology shape these two distinct ways of knowing the salmon.

Third, Berger and Luckmann sought to develop an approach that would reveal the processes by which reality, knowledge, and meaning develop in societies. In so doing they grappled with one of the great divisions in sociology: the micro-macro divide. These two levels of society seem paradoxically distinct from one another. On one hand, we find ourselves living lives that put us into daily contact with friends, family, coworkers, and strangers: microsocial interactions. Yet we also live against a backdrop of "social facts," including macrosocial institutions like education, government, and the economy. These are very real to us, even though we cannot point to them. Try finding "government." You may identify a building or an elected representative, but those are only tiny parts of government, not government itself. In many ways its presence in our lives is not material, but it is nevertheless real.

We live with both realities, the micro and the macro. But that begs the question of how they are connected. Berger and Luckmann recognized that through the course of time some microlevel interactions evolve into macro, structural institutions. That is, some *subjective* meanings that initially are embraced by a small group become institutions: *objective* facts taken for granted as social forces that have a presence in everyone's lives. In other words, during the course of history all social institutions emerge from ideas first generated through microlevel, small-group interactions.

In hindsight it can be seen how these institutions emerge. One straightforward example is the United States of America. Prior to 1776 it did not exist as an institution, of course. However, beginning with the interactions among a small group of discontented British colonists sometime before that, the notion of independence grew. Even with the signing of the Declaration of Independence, the institution that we now think of as the "United States" was nonexistent. The United States of America was a highly personal, subjective concept. Only through the course of the Revolutionary War, the mustering of the bureaucratic power of the United States government (however tentative and weak), the adoption of the Constitution in 1789, and the events of the following decades and centuries did the United States as an institution—something huge and distant from us, yet real—begin to take shape. Beginning with the grumblings of a few angry dissidents, the *United States of America* evolved from a microlevel, subjective vision to an objective social fact.

Fourth, there is a danger to this objective meaning-making, however: *reification*. Berger and Luckmann explain, writing that "reification is the apprehension of the products of human activity *as if* they were something else than human products—such as facts of nature, results of cosmic laws, or manifestations of divine will. Reification implies that man [*sic*] is capable of forgetting his [*sic*] own authorship of the human world. . . . The reified world is, by definition, a dehumanized world."[16] To avoid reification, constructivism directs us to own up to the world as of our own making. *Everything* has meaning. If we can speak of it, whatever *it* is, that thing has meaning and so has been touched by society. The danger comes when we treat *concepts* like "government" or "nature" as if they were real things. The concept—the word—is not the thing. I am reminded of the saying, "The map is not the land." Maps and words are representations. They are not reality. But because these representations or symbols are imbued with meaning, they have an immense power in our lives.

This book is an exploration of how salmon have come to be treated as a reified social fact by scientists and how others are struggling at the microlevel of society—through their interactions in small groups—to change those meanings. As David Ralph Matthews wrote, "Viewed from this phenomenological or constructionist perspective, the meaning of nature is not a constant, but varies in terms of the frame of reference we bring to it."[17] As we explore this social creation called Nature, I think we will discover a great deal about ourselves and about our society.

MACROCONSTRUCTIONS AND RATIONALITY

Implied in this discussion is the notion that constructions are subject to change and that they may differ tremendously, even among groups in the same society. For the most part, that is correct. However, some social constructions seem to have a persistence that nearly all others lack. These *macroconstructions* underlie so much of social existence that they become reified in the most extreme of senses. I have a friend who meditates, and once when we were talking about his mantra I asked him what it meant. He paused, thinking, then smiled. "Everything," he said. Such is the case with macroconstructions. They are everywhere, it seems, society's reified mantra. We find ourselves accepting their reality as if their existence were so certain and permanent that they are on a different plane from us. They are cornerstone principles of social life, meanings that define who we are as a society, yet they are rarely if ever acknowledged.

No macroconstruction is more powerful in our society than *rationality*. Max Weber, a German sociologist who wrote in the early twentieth century, referred to several different types of rationality,[18] but his concept of *formal* rationality is what interests us here. By formal rationality Weber meant activity that is goal-oriented, calculated, and undertaken with reference to established rules, laws, or procedures, like the step-by-step operations of a fast-food restaurant, where even the greetings to customers are specified by the corporation. Weber argued that in industrial society, rational impulses creep into all aspects of social life. They influence the practice and the meaning of work, family, religion, and all else. Thus my labeling of rationality as a macroconstruction: rationality has left no part of modern society untouched, leading George Ritzer to refer to rationality as the "McDonaldization of society."[19]

Rationality as pursued through economic, governmental, legal, bureaucratic, and scientific endeavors is omnipresent in capitalist industrial societies. These rationalizing forces are distinguished by their ceaseless promotion of and emphasis on efficiency, predictability, calculability, and control. We will see that these characteristics are found throughout salmon biology, and they strongly influence the dominant biological construction of salmon.

However, Weber saw that rationality has its limits. Control and power, central to the rational impulse, are the subject of Chapters Two, Three, and Four. We will see salmon and salmon biologists directed by society's most powerful controlling forces. Ultimately, though, those forces clash with others that Weber, at his most pessimistic, felt were certain to be permanently lost as rationality grew: freedom and what I call "self-determination."

For salmon this means that there is hope, for even today, at the moment when control and power over salmon are at their height, a movement has begun among biologists to de-rationalize salmon biology and thereby to de-rationalize the fish, as I discuss in Chapters Five and Six. This movement is only now gaining momentum, and its ultimate impact is uncertain at best, given that it clashes not only with a rationalized scientific paradigm that has dominated North American fisheries studies for more than a century, but with the rationality macroconstruction itself. De-rationalizing salmon is an effort akin to the courageous, tortuous struggle of salmon against the massive bulk of a dam blocking their way. But just as dams have fish ladders to allow the salmon to make it upstream, we will see that society's own values and institutions—the same ones that promote rationality—also allow a minority of biologists to fight against the flood.

RATIONALIZATION AND THE SOCIAL CONSTRUCTION OF NATURE

No explicit link exists between Weber's rationality concept and Nature. However, related themes—such as controlling Nature through science and technology—may be found in Weber's work, as when he wrote,

> Now, the peculiar form of Western capitalism has been, at first sight, strongly influenced by the development of technological possibilities. Its rationality is today essentially dependent on the calculability of the most important technical factors. But this means fundamentally that it is dependent on the peculiarities of modern science, especially the natural sciences based on mathematics and exact and rational experiment. On the other hand, the development of these sciences and of the technique [technology] resting upon them now receives important stimulation from these capitalistic interests in its practical economic application.[20]

Weber's logic regarding the edifice of all modern rationality was the following: bureaucratic, governmental, and economic forces rest atop a scientific and technological layer of production; this in turn is dependent upon the presence of a Nature that is accessible, available for use. Capitalist economics, hierarchical organization, science, and politics/government—simultaneously the embodiments of rationality and the forces through which formal, goal-oriented rationality is extended into all walks of life—have as their aim the efficiency-producing ordering of social life. Endless, boundless Nature emerges as foundational to our industrial society.

It appears that Weber was cognizant of the importance of Nature to the industrial mode of production. He wrote, "This order is now bound to the technical and economic conditions of machine production which to-day determine the lives of all the individuals who are born into this mechanism, not only those directly concerned with economic acquisition, with irresistible force. Perhaps it will so determine them until the last ton of fossilized coal is burnt."[21] Undoubtedly, Weber was commenting on the human cruelty inherent in industrialism. But he was also well aware that without "natural resources" like coal the industrial social "mechanism" grinds to a halt.[22]

Like human labor, coal and other "resources" are vitally important for industrial production. And they are essential not only for industrial processes, but for scientific endeavors as well. Without iron ore, steel cannot be milled; without water, fields cannot be irrigated; and without salmon, salmon biology cannot proceed, nor can salmon fishing or any other scientific, economic, or technological process depending upon the salmon. These taken-for-granteds indicate that life in industrial society cannot exist without this Natural foundation. This holds for all societies. Gatherer-hunter societies forage for food, make weapons from carefully crafted wood and stone, and burn trees for warmth. But although all societies use the stuff of the planet, modern society is marked by its rationalistic approach to the Natural world. Rather than conceived as merely apart from society, Nature has been reconstructed as something foreign to be efficiently and productively used for narrowly human ends through science's theories and calculations.

Storytelling

This book is similar to a history or even a work of fiction, for the object here is to tell a good, complete story. The tale to be told here is the biologists' constructions of the salmon, although I must acknowledge at the outset that this story is but one version of the reality that salmon and salmon biologists experience. I think it is reasonably true to salmon biologists' reality (I would not presume to speak for the salmon). I say this both because I have spent a lot of time with biologists and I think I have a fair idea of their world, and because I have made presentations of my research at which biologists were present. Their feedback was never negative, and in some cases it was enthusiastically supportive. But no doubt some biologists will take issue with this version of their world, or at least with parts of it.

In a nutshell, here is how this version of the story unfolds. Chapters Two and Three are complementary. Both explore the concept of "control" from different angles. Chapter Two details how social institutions limit biologists' behaviors and constrain the constructions of salmon available to the scientists. Chapter Three examines two other aspects of control that are central to salmon biology: the adoption by these scientists of an engineering model for understanding salmon and the efforts of salmon biologists to extend their control over other social groups, in particular their chief antagonists, "managers." In Chapter Four, control is exemplified by "fish culture," the marine version of agriculture, as practiced in salmon hatcheries. In hatcheries we see how the fish are both physically and cognitively constructed through their technologization and through the ways that biologists speak of them.

By this point, it will be clear that salmon biology as a social practice operates not much differently than a lot of other social groups do, with frequent internecine squabbles, too little money to do all that needs doing, and resistance to external authority, to name but a few of the commonalities it shares with many professions. In the process, a certain salmon is produced, and it may not be the one that either Nature or biology intended. Chapter Five explores this uncertainty by taking an extreme position: treating salmon biology as "myth," just one of the many possible ways of telling the salmon tale. Salmon biology is fraught with uncertainty. When the unknowns pile up while, simultaneously, there is tremendous pressure to provide society with answers, perhaps it is understandable that the sheen and power of science becomes a bit tarnished. Some biologists have reacted to the pressure that they and the dwindling salmon runs are under by questioning the fundamental social assumptions of their profession, as we shall see in Chapter Six. In contrast to control, these biologists seek greater freedom for themselves as researchers and for the salmon that they study. Some have even taken on the role of advocate, championing the salmon cause in ways that only environmentalists had done before them.

Chapter Seven extends some of the themes from the preceding chapters, applying them to an on-going confrontation between the United States and Canada. On the surface this "Salmon War" appears to be over which nation's fish are whose, but it is really a fight over control, freedom, and the meaning of community. In Chapter Eight I summarize the book's key insights and also acknowledge that an "unconstructed" nature is possible, even if only to us individually. For those interested in the methods used to gather and analyze the data, or who would like more information on other works on the social construction of Nature, I have included an Appendix covering these topics.

In terms of intellectual traditions, *Fishy Business* has a foot in at least two camps. The primary one is "environmental sociology," which is best understood as the study of society's interactions with the nonhuman world. Environmental sociologists' studies cover a tremendous range of topics, including attitudes toward the Environment, environmental racism, environmental policy, the economics of resource exploitation, and, increasingly, the social construction of Nature. From its inception it has challenged sociology to look beyond the narrowly social and to explore the impacts of culture on ecology. Yet in its excitement and urgency, environmental sociology has been slow to explore the extent to which Nature and the Environment are themselves social products. In a sense, then, I see this book as examining a conceptual foundation that has long been taken for granted.

The other primary area of study that has influenced my approach to this topic is the sociology of science, especially the social construction of science and technology. This is a controversial field, one that some scientists have reacted to quite harshly. What troubles them most is the notion embraced by some, but not all, writers in this area that science may not merely transmit knowledge from Nature but that it both creates the knowledge and, worse yet, the knowledge is entirely a social product. Fortunately, it is possible to reject this "strong" version of the social construction of science while still seeing the touch of society (rather than its heavy hand) in the outcomes of scientific endeavors.

AN AUTHOR'S STORY

Before we embark on this tale, a word about myself as author is due. This is in keeping with the constructivist and ethnographic approach taken here, which directs that authors accept responsibility for their ideas by acknowledging their outlooks and biases (a process known as "self-reflexivity"). My viewpoints are strongly felt. Industrial society has abused the planet in the worst of ways, and few other societies in world history have shown so little concern for the place that sustains them. The dams and overfishing that I write about are reprehensible products of a culture that cannot see beyond the tip of its nose and a people who are foolish enough to believe that science and technology alone will save us. That said, in the pages that follow I hope that I do justice to those with whom I disagree. Objectivity is a chimera, but fairness is the least I owe to the biologists who generously shared their lives, ideas, and opinions with me.

2 Who—or What—Is in Control Here?

JIM WINSTON IS A university biologist with a small but active stable of graduate students who assist him with his salmon research.* An American with decades of experience behind him, he was one of several scientists with whom I spoke who have witnessed dramatic changes in the interactions between society and salmon. His spacious and meticulously well-kept office was replete with photographs—some of them years old, judging by their faded colors—of Winston holding large salmon that he had caught on vacations. I was left with the impression that salmon were more than a vocation to him. They were the center of his professional and private lives alike.

As we spoke he justified this impression. At one point he almost gently vented his anger about the ways many states *manage* salmon and other fish: how states have controlled fish for human purposes. This was prompted when I asked about the practice of "stocking" lakes, a management technique whereby the state places fish reared in hatcheries into lakes and streams that, in many cases, have never had fish living there or have not been home to the species of fish being introduced. Stocking is a time-honored approach to meeting the demands of anglers, but one grossly out of touch with contemporary science, said Winston, who spoke forcefully yet never altered his facial expression:

> Now that is a legitimate management technique for waters that for some reason have been denuded of any important fish species to the sporting public. But that can't be the management philosophy, not only state-wide but in the whole Pacific Northwest. It's just simply wrong. It is out of synchrony with the behavior and the suitability

*I have given all of the biologists who participated in this project pseudonyms, and other characteristics about some of them or about their locales also have been altered to ensure their anonymity.

of the environment with respect to the species that they are introducing. It's out of synchrony with what the ecology of the system has gone through to accommodate the different species. But you see that philosophy promoted because it's what the public wants in its fishery system. So 80 percent of fishery biologists are involved in those types of activities that are out of synchrony with what the system should really have.

* * *

Taking a brief break from the harried annual conference of the American Fisheries Society's Washington–British Columbia chapter, Mark Johnson spoke with a confident intensity. He was eating lunch while we talked, but I never had the feeling that he needed the mouthfuls of a French dip sandwich to give him time to think about his answers to my questions. Johnson's manner bespoke someone comfortable with his position, that of a veteran biologist with the Canadian Department of Fisheries and Oceans who knew his work and cared a great deal about it.

My final question to him, and to most of those who participated in this study, was, "What is *good biology* to you?" Many of the biologists paused or laughed or said something to the effect of, "You should have warned me you were going to ask that!" They seemed intimidated. But not Johnson. He swallowed a bite and said, "Good biology means not only can you make predictions about what's going to happen, but you can also identify the mechanisms that support those predictions so that you have true insight into why, when Condition A changes, Outcome B occurs. You know what the mechanism is, you know what the linkage is. When you have that kind of insight and you have that kind of knowledgeability, it is much less likely that you will be rudely surprised by large departures in expected performance."

* * *

These quotes by Winston and Johnson demonstrate the extent and character of a central concept emerging from my research: the role of control in salmon biology. Control, one of the cornerstones of our rationalized society, is a "taken-for-granted" for salmon biologists. That is, their work is so much about control that as a general theme it goes unnoticed and unquestioned. Few biologists ever used the word control in our discussions, but its presence was unmistakable. Winston, for example, did not argue that the control practices inherent in "management" were unacceptable, only that particular kinds of control were inappropriate, such as "bucket biology," when fish are thoughtlessly dumped into a lake or stream where they do not belong. Similarly, Johnson's stress on "predictions" exhibits the control over Nature's uncertainties that science prizes.

For these and all other salmon biologists, control in numerous incarnations is ubiquitous. The particulars of control, such as the best management approaches (the best ways to manipulate salmon for society's benefit), are open to debate, but whether to exert control is not. What salmon biologists control is varied; moreover—and this is as much taken for granted by many biologists as whether to control—biologists themselves are controlled. Their work is constrained by rationalizing political/governmental and economic forces and by an array of other social institutions.

In the following sections I briefly outline the place of control in salmon biology. Then I note the historical basis for that control, found in the emergence and development of the fisheries profession. Finally, the bulk of the chapter is spent exploring how politics and economics shape salmon biologists' work.

Control, Power, and Salmon Biology

One of the things that scholars do better than anyone else, other than politicians, is redefining ordinary concepts. The worst of these efforts confuse issues and obscure key points; the best of them illuminate those points like the sun, giving us a tool with which we can poke and prod the world, opening the doors of understanding. Jack P. Gibbs achieved the latter when he argued that a careful distinction between "control" and "power" should be made by scholars. Boiled down to its core, Gibbs said that control is the actions that a person or a group takes in an attempt to direct some part of its world. Power is the mental, cognitive counterpart to control; it amounts to knowing or believing that control is possible.[1] An example of control is the act of catching a salmon (whether by net or fly line is immaterial); power is the mental idea that such a feat is possible—that control in the form of catching a salmon may be achieved.

Control manifests itself in three ways in salmon biology, and each of the three was exemplified in the quotes that opened this chapter. *Structural control over salmon biology* is pursued by macrosocial political/governmental and economic entities. An example of this was Jim Winston's complaint about the role of fisheries biologists—including salmon biologists—in ecologically inappropriate management practices. Roles are defined by powerful, controlling organizations like state fish and game departments. Many fisheries biologists work at the behest of these politically oriented agencies that carry out policies governing things like which species of fish should be used to stock particular lakes and streams.

Politics is an especially potent controlling force because the departments are largely supported by license fees paid by hunters and anglers, not by general tax dollars. The effect of funding fish and game commissions and departments in this way is to vest the hunting and fishing interest groups in the Pacific Northwest states with substantial political and economic power, because their members are the ones footing the bill for fisheries projects and research. This means that not just anyone's voice will be heard when it comes to wildlife management. Winston objected to biologically unsophisticated, and often ecologically disastrous, programs that have resulted in popular but non-native, so-called exotic game fish being planted in areas where they have decimated local, native fish populations, including salmon. He was angered that biologists have agreed to carry out such destructive activities, but what choice did they have? As we shall see, political and economic control over salmon biology occurs in even more direct ways.

In Chapter Three I discuss the other two aspects of control that are of interest to us. The first of these, *control of salmon by salmon biologists,* is indispensable to the pursuit of salmon biology. Such control is achieved in numerous ways, including conducting laboratory experiments on salmon, imposing an engineering-like "systems" view on salmon and their environments, and quantifying the fish and their surroundings—counting the salmon, collecting other data, and then creating mathematical models that predict the system's behavior. Winston's emphasis on "the ecology of the system" and the "fishery system" and Mark Johnson's use of the word "mechanism" reflect the dominant place of engineering and mechanical metaphors among salmon biologists (and biologists in general). Moreover, Johnson implies that traditional scientific studies leading to "knowledgeability"—prediction, certainty, and perhaps even behavioral laws for salmon—are the ultimate aim of "good biology." The further implication is that control over salmon is possible and that it is a worthy goal.

Control over other social groups, the third component of the control concept, also was exhibited in these quotes. In his obvious disdain of "management" as it usually is pursued, Winston voiced a common theme among salmon biologists: their dispute with others over who best knows how society should interact with salmon. Especially noteworthy is the control that salmon biologists seek over the agency managers who make the ultimate decisions about things like stocking lakes and fishing regulations.

Professionalizing Biologists and Salmon: A Brief History

Control's place in salmon biologists' lives is best understood by considering the history of the fisheries biology field, because control is a *process*. It emerges through time. What stands out most sharply in the discussion that follows is the indisputable place of economics and politics in the development of fisheries science. Economics dominated the rationale behind the establishment of the continent's foremost fisheries biology organization, the American Fisheries Society (AFS), and other powerful rationalizing influences—bureaucratization and the politicization of the fisheries profession—closely followed economics in their impact.

In 1870 the American Fish Culturists' Association, precursor of the AFS, was founded. The very name, *Culturists'*, highlighted the economic overtones of the organization. As the author of the official AFS history of its early decades wrote, during the years immediately following the Civil War "observers became aware of a decline in the populations of certain valuable species, particularly salmon and trout, in many eastern lakes and streams.... Under the circumstances, it was natural for the fish culturists and others concerned about the resource to want to replenish the depleted waters with their own hatchery-reared trout and salmon."[2] Fish culturists had strong economic interests in fish, and the fish—whatever the species might be—were seen as something to be cultivated; they were an *agricultural* commodity to be grown for food and profit. This is a telling theme, one that persists today in the discipline, not only in fish hatcheries that culture fish in mass quantities, but in how biologists view wild fish populations as well.

The fish culturists quickly recognized that the federal government had a role to play in furthering their interests. In 1871 Congress established the Commission on Fish and Fisheries (later the U.S. Bureau of Fisheries and now the Fish and Wildlife Service). The following year "the Association voted to ask the United States Congress to take part in the 'great undertaking' of introducing or multiplying shad, salmon, or other valuable fish throughout the country, especially in waters under federal jurisdiction or in interstate and boundary waters."[3] Fisheries advocates recognized that it would benefit them to tap into the government's deep pockets. Thus began a lucrative relationship that lasts to this day, though the money now goes not to culturists but to scientists. The federal government began what today would be called "enhancement," "supplementation," and "introductions" of fish, all of which entail extensive control by manipulating fish to serve human ends.

Science was a crucial component in these efforts, as evidenced by Association President Robert B. Roosevelt's remarks at the group's annual meeting in 1876: "It had been shown by the able and scientific labors of the United States Commissioner, Mr. Baird, that there need be no fear of scarcity of fish food either in the ocean or in our great lakes. . . . A new science was being born into the world, but the clear light is visible at last. There need be no fear for the future, and in much less than a hundred years the waters of America will teem with food for the poor and hungry, which all may come and take."[4] The nascent science of fisheries would feed America by coaxing a sea of plenty from the nation's waters. And when the association's name was changed to the American Fisheries Society in 1885 and its constitution rewritten, its first article directed that the society's "object shall be . . . the treatment of all questions regarding fish, of a scientific and economic character."[5] Fisheries science was firmly, officially wedded to economics.

The relationship was not without some strife, however. In its early years there emerged within the AFS a tension between scientifically oriented fisheries "enthusiasts" and the founding "hatcherymen," whose orientation was predominantly economic.[6] The tension declined slowly as "growth and segmentation in the field of aquatic biology" increased through the years and as fish biology in the United States was rationalized; powerful social interests identified scientific control over fish as the most efficient means to realizing economic benefits from the "fish resource."[7]

Clearly, science needed to be involved for capitalism to control fish. And so it was. The Woods Hole Marine Biological Laboratory in Massachusetts was established as the first government-run center for fisheries research in the United States in 1880; a second center was founded in North Carolina in 1901. Canada's first such institution was created in 1898, and numerous state and provincial laboratories opened in the 1920s and later.[8] Especially in Canada, government laboratories have long been the centers of scientific research. In the United States the leading research locales have been at universities for sixty-five years or more.

SCHOOLS OF FISHERIES

The first school of fisheries was launched at the University of Washington in 1919, and today it remains the foremost center for fisheries research north of San Diego. Washington and the other fisheries schools began not only as efforts to bring more science to bear on fisheries questions, but to "train fisheries professionals"—to socialize students by inculcating in them accepted scientific methods, and thereby to direct their approach to fisheries science and management.[9] Education was to play an essential role in the rationalization of fish.

In a letter written in 1914 to the University of Washington president, the U.S. Fisheries commissioner commented that "a school of fisheries, or at least a comprehensive course in fisheries, [would have] for its object the equipping of young men and young women for practical work in the service of the federal government, the various states, and private establishments having to do with artificial propagation, the curing and marketing of fishery products, and the administration of the fishing industry."[10] The Pacific Fisheries Society pledged its support for the concept, but the idea languished for five years until the College of Fisheries, offering courses toward the bachelor of science and master of science degrees, was established under the directorship of John N. Cobb, a former assistant general superintendent of the Alaska Packers Association, a fish-processing company.[11] The emphasis on fish economics and management in those early years is demonstrated not only by Cobb's background, but by the titles of some of the courses: "Pacific Fisheries," "Fishing Vessels and Boats," "Fish Culture," "Fishery Methods," "Fresh and Frozen Fishery Products," "Curing of Fishery Products," "Fundamentals of Canning," "Canning Machinery and Cannery Management," "Canning of Fishery Products," and "Exploration of the Sea and its Relation to Economic Food Fishes."[12]

In 1935, after considerable turmoil, including the disbanding of the College of Fisheries, the reorganized School of Fisheries was established. Although the faculty and students of the school long have been engaged in "pure," "basic," "fundamental" research, and the current course listings have a far less (directly) economic orientation to them than those of earlier years, the essential idea behind the concept *fisheries* largely remains one of *applied* science in the interest of economic exploitation, putting science to work whether it be for commercial or recreational purposes.

Fisheries is characterized by a rationalistic, utilitarian ethic. Historically, and to a large extent even today, fisheries biology has been about making money—or helping others to do so. One recent graduate of the University of Washington School of Fisheries whose prior background was in zoology put it to me this way:

> I rather naively mistook "fisheries" for "fish biology." . . . [I]t was a very, very big difference in the sense of a limited variety of fishes that they even were looking at and the limited kinds of questions in which they were investigating ecosystems.
>
> Sort of the typical fisheries situation is you are given the species and the habitat, and then you try to find the most useful thing that you can do with that species in that lake, whereas a zoologist sits back and thinks about the question and says, "Okay, what is the most interesting species in which to pursue this question and where is the most appropriate place to?" or something like that. I kind of made my way, but it was a bit of an adjustment. I think it's a bit easier now that the program has diversified some more.

Similarly, another University of Washington graduate, also with a zoology background, observed,

> I think it's interesting to look up in British Columbia, where fisheries, for the most part, is within the Department of Zoology, from what I understand. . . . They've done a lot of important fisheries work in other places, but often it has been in zoology departments. Maybe that has some influence on how they've looked at things. I think they've done a better job of conserving their resources, by and large, than the Americans have, and I doubt that you could attribute the whole thing to something like that. But I think maybe they've had a slightly broader perspective on things—how the salmon fits into the total system—more so than the Americans have, historically.

Today, as throughout its history, the fisheries approach is driven by control as framed by economic and political contingencies. Fisheries biology predominantly emphasizes the things that matter to those rationalizing social forces.

Structural Control over Salmon Biology

The Political and Economic Milieu

Economics and politics always have been fisheries biology's raison d'être as a profession. They pervade salmon biology in particular in myriad ways, and I was struck by the extent of awareness among biologists of the political and economic context in which they undertake their work. That salmon are a much-discussed social issue is readily apparent to anyone in the Pacific Northwest who has an interest in public affairs; the pervasiveness of salmon in the news testifies to this.[13] Yet the biologists' knowledge of and concern for the macrolevel, political/governmental and economic climate as it relates to salmon initially surprised me. Even when I came to understand that the biologists perceived their destiny (professionally and, for a minority, emotionally and even spiritually) as closely linked to that of the salmon, the biologists' attention to the details of relevant political, policy, and economic developments continued to impress me.

However, some biologists are more affected by macrostructural limitations than others. For example, a hatchery biologist observed that political infighting over the future of wild fish in the region was wearying: "It's difficult sometimes. Every once in a while you have to remind them that we're working for the same thing, trying to improve the resource. All the time there's more people with what I call an oar in the water here. We got more committees, more groups, whatever. It makes it difficult to coordinate everything." Salmon biologists are not passively embedded in a social structure, however. Increasingly they attempt to affect the social forces that direct their work. At times they are brought into the political discourse in a deliberate manner, such as when they are named to advisory committees like those considering salmon recovery plans mandated by the listing of some runs under the Endangered Species Act (ESA). In other instances the involvement by biologists is more assertive, even aggressive, as when they file petitions with the federal government to list runs under the ESA (see Chapter Six, below).[14] However, the opportunity to participate on panels and to advocate for the fish has been granted by the government; it is not an inherent right by virtue of citizenship or membership in a scientific elite. In itself, this is a straightforward example of the ways that structural forces exert social control over salmon biology.

Whether they are employed by governments, by corporations, or as consultants, biologists' work is made possible by economic and political momentum or, perhaps, by economic and political expediency, not by some mythical "scientific imperative." Stephan Fuchs addressed this theme when he wrote, "The sociologist of scientific knowledge . . . points out that science reacts to social pressures—just like any other ordinary and mundane system of action. In this view, science is like art or politics and cannot claim special epistemic status."[15] Fuchs's words echo those of Berger and Luckmann regarding the relativity of knowledge. Like all other ways of knowing, science is not special, unique, or different in a sociological sense, something recognized long ago by Robert Merton, a pioneer in social studies of science.[16] Science is strongly influenced by social forces.

Such observations disturb and even anger many in the sciences.[17] But the data from this and other studies do not support the commonplace belief that science is above the social fray. Consider the following comment by Gary Hite, a gregarious and obviously brilliant Canadian biologist whose specialty is developing computerized models for government agencies:

> More and more, especially in B.C. [British Columbia] and especially in the Department of Fisheries and Oceans here in B.C., people are getting, because they've been told to do so, more and more involved in comanagement with native [aboriginal] people. The whole thing has turned around. Instead of scientists basically being the managers, too, and saying, "This is what you have to do," it's basically turned around and saying, "Well, the native groups have taken control of a lot of these things. How can we help you do what you need to do?" So I see a real change here in B.C. in the last year. DFO [Canada's Department of Fisheries and Oceans] has made a really strong commitment to comanagement and working with native people. Up until a couple of years ago there was very much a strongly antagonistic position between native people and the managers and the scientists. And then all of a sudden from up the political ladder people said, "A lot of money is going to go to comanagement." So now if you want to do work, you have to do it in that area, which is a real turnaround.

That native peoples are receiving attention from the Canadian government may be less an indication of the government's heartfelt concern for aboriginal rights than it is an effort to head off a divisive confrontation akin to what occurred in the 1970s in Washington State. This is an example of political expediency, with scientists, a government "resource," treated as part of a quid pro quo arrangement whereby professional expertise is offered in exchange for a nonconfrontational approach to addressing aboriginal fishing rights.

In contrast, in Washington during the 1970s the "salmon wars" between the state and indigenous Native American tribes were settled in federal court in what is known as the "Boldt decision."[18] It left a profound political legacy, one that savvy biologists take into account as they consider the future of salmon and of their work. A biologist familiar with efforts to revive dwindling chinook salmon populations in Idaho said with an ironic tone, "In fact, if we actually did a good job of bringing back fall chinook [listed as a threatened species in

Idaho], it probably just exacerbates the problem for the State of Idaho because we would probably fish harder down in the Zone 6 area between Bonneville [Dam] upstream to McNary [Dam, on the Columbia River between Oregon and Washington] with the Indian net fishermen and harvest more fall chinook and also harvest more steelhead at the same time. . . . Those are kind of little public policy factors that weigh into the bargain in what's going on." In other words, efforts undertaken in one area of the massive river system to help salmon might be for naught because Native American tribes with treaty fishing rights could catch the additional fish destined for streams far upriver. Political realities surrounding the preservation of salmon are so complex that decisions intended to preserve salmon in one jurisdiction may actually backfire and create new problems for salmon, and thus for social groups as diverse as environmentalists and foresters, in other areas.

Salmon biologists recognize this, and many factor it into their endeavors. What struck me was that regardless of the extent of their involvement with government policy or with economic enterprises, *all* of the salmon biologists displayed a substantial amount of political and social sophistication. They were aware of the myriad issues surrounding salmon, whether or not those issues directly affected their work. In this, they all were "expert 'sociologists' "[19] and talented social actors, *salmon biologist-sociologists* akin to Michel Callon's "engineer-sociologists."

Callon wrote of the engineers he studied, "Whether they want to or not, they are transformed into sociologists."[20] The same is true for salmon biologists. As social actors they are embedded in a social structure, and to be effective in their discipline they must familiarize themselves with the social forces all around them. Scientists are embedded because they work deeply connected to issues of concern to the broader social world. In his study of the sociology of economic activity, Mark Granovetter noted the importance of the "[s]tructural embeddedness" of individuals. One must be aware of her or his social position in relation to others and in relation to macrostructural forces in order to effectively negotiate the details of daily life.[21] The other side of embeddedness is its importance to powerful social forces because it "facilitate[s] social control."[22] In the case of salmon biologists, they find themselves embedded in a social structure dominated by governmental policy makers, politicians, bureaucratic agencies, and corporations. Governments and corporations direct biologists' activities to serve public policy and private interests, in the process exerting substantial control over biologists' constructions of salmon.

FUNDING SALMON BIOLOGY

No social group operates independently of controlling social forces, and for salmon biologists this is most clear in the funding of their research. Funding is of utmost importance to most of the biologists with whom I spoke. Without money for laboratory equipment, boats, nets, computers, travel, and countless other things, salmon biology would be almost nonexistent. The enterprise is simply too expensive for researchers, and even for most government agencies and universities, to undertake on their own. Biologists' success in obtaining funding depends in large part on the political and economic importance of salmon, as I show below. I also examine the extent of external control over salmon research and biologists' reaction to it, including the politics of funding within the salmon biology profession and the need to "bootleg" research. Throughout this discussion I draw comparisons between Canadian and U.S. approaches to funding and their implications for salmon biology and for constructing salmon.

Society and Funding for Salmon Research

Salmon probably are the most studied fish in the world, yet much is not known about them. Nevertheless, biologists often make their cases for more salmon research based on the fish's economic value, not "scientific curiosity" or the exploration of the unknown. One of my research participants (interviewees) told me that

> there's lots of interest in the stock differentiation of salmon [distinguishing different populations of salmon to help commercial fleets avoid wiping out smaller populations, or "stocks"], so that creates a financial basis for this type of question. That wasn't a driving factor at all in doing research, but the money was. . . . When I said that my Ph.D. research was really not financially based and there was

very little interest in sockeye-kokanee, the interest in sockeye and kokanee has since exploded. And it's exploded because of sockeye salmon as an endangered species. . . . Salmon had many of the same elements as whitefish did, and salmon are closely related to whitefish, same family. There's a lot more money in salmon research than in whitefish research, so that's why I went to study salmon.

Biologists study the "money fish" because those economically valuable species attract research funding. At the dock the commercial salmon catch is worth hundreds of millions of dollars annually to the United States and Canada alike.[23] As one indicator of the close ties between science and economics, consider one biologist who heads a group that for decades has been awarded annual grants, each worth several hundred thousand dollars, from the Alaska salmon industry. The researchers provide information on salmon numbers and location needed by the industry, and in return the scientists have the opportunity to conduct research that has led to numerous publications. Greater funding and publishing opportunities increase individual biologists' standing within the biology community. Economically and politically important, salmon become constructed as vehicles for status attainment.

Society's overwhelming interest in the salmon, whether economic, political, or legal (as found in laws such as the ESA and the Northwest Power Planning Act),[24] has produced huge sums of money for research, and careers can be made by studying these species. David Perkins, a young biologist whose name was mentioned in numerous interviews as a "rising star" in the field, noted that the primary source for research funds for American researchers in the region, not only university researchers but those in government agencies as well, is the Bonneville Power Administration (BPA). BPA is a federal agency that sells the electricity produced by twenty-nine of the largest hydroelectric dams in the region. Those dams have been implicated in the precipitous decline in salmon numbers over the last six decades, and BPA looks to science to help it revive the fish runs. My interchange with Perkins went like this:

PERKINS: The deep pocket around here in fisheries is Bonneville Power Administration. That's *the* deep pocket to the extent that. . . . Well, it's like their research-enhancement-mitigation budget is close to a hundred million a year. Huge. They fund a lot of good stuff and some not so good stuff. It's kind of a funny situation to have so much of the funding concentrated in that area. You have the federal government deriving a lot of their money from BPA. It's kind of a weird deal. You have state, federal, Indian tribes, consulting companies, universities all applying to this very deep pocket.

Q. That's something that I wasn't aware of. The federal government agencies, NMFS, for example, will go to BPA for funding?

PERKINS: Yes! The National Marine Fisheries Service has a . . . division, they've changed their name—I think they're called "Coastal Zone and Estuarine Studies Division." Does basically Columbia River stuff. I was at their review two years ago or so. They said 75 percent of their money is non-NOAA funds. Seventy-five percent. By far the largest portion is BPA and then Army Corps of Engineers. So they're heavily, heavily dependent on extramural funds. If they only relied on NMFS funds, they'd have to cut back on the order of 75 percent. They also get money from U.S.-Canada [Treaty] because a lot of the genetics work is work that they're figuring out, identifying salmon populations in mixed-stock fisheries, so they get a lot of U.S.-Canada money, too. Washington Department of Fisheries gets lots of BPA money for genetics for scale otolith—fish ear bones, used to determine the fish's age—work, of course a lot of hatchery stuff. Indian tribes get lots of it. So it spreads pretty wide.

Despite these deep pockets, researchers complain that there is not enough money to fund all of the projects that deserve support. More than anything else, the lack of funding constrains biologists' work. Put simply, *funding is the crucial control on salmon biologists' research.* Canadian Klaus Huberman, nearing retirement after a distinguished thirty-year career, put it to me this way: "Any question can be asked if you can get funding for it! That's a separate question."

All fields of study experience funding constraints. The deeper issue is which projects receive funding and why. This grants-making process rewards certain lines of research and penalizes those researchers whose interests stray too far from accepted and expected topics and methods. So the question is not exclusively one of economics and quantity—of dollar availability for research projects—but one of politics and quality as well—dollar availability for *particular* projects. Funding is an expression of social control, especially the power of the agencies that identify areas worthy of research support and that distribute research dollars. That control affects not only what salmon biologists do but the knowledge that they create as well. The topics of interest to biologists are dictated. This limits the range of questions they ask and the insights that they produce.

Even for successful researchers whose work is well within these normative boundaries, obtaining research funding is an anxiety-producing proposition. Take, for example, the following interchange with a respected geneticist, Alex Stand:

> *Q. Is that a problem, as of late, to get funding?*
>
> STAND: In general it is. I think we've been fortunate recently in that we've been able to keep the research going, but I always have to be thinking about getting in a new grant proposal and figure out how to keep it competitive and so on.
>
> *Q. Does the lack of funding serve as a barrier to research? Would that be safe to say?*
>
> STAND: I think all researchers would like there to be more money for research, obviously. When I look at the total situation in the country, with as many problems as we've got, I think all in all the research funding situation is not that bad. I think that there's still funding available if people are willing to be creative about how they approach things and to dig in there and work hard at it. I think, yeah, we could use some more research funding, but on the other hand I don't think that, relative to the total situation, I don't know that we deserve to complain that much.

Researchers were loath to characterize the lack of adequate funding as a "barrier." For the most part they were only slightly more willing to admit that their freedom to research whatever they wished was constrained by the needs or demands of funding agencies. Jim Winston said that BPA "probably influences the proposals that the scientists develop. But hard times are hard times. BPA is a good funding source, so universities will go to BPA, and what do they go to BPA with? The projects that will answer the questions that BPA and the Power Planning Council are concerned about. So the criticalness of the issue is not as important a driving force as the availability of dollars to do the research." Researchers, especially those in academic settings, often assert that they are free to explore what they will. The realities of a highly competitive research milieu, however, do control the questions that biologists may ask and the methods that they may use to address those questions.

David Perkins remarked on the reality that the piper calls the tune in the Columbia-Snake region regarding which lines of research are funded: "Oh, I don't think there's any question of that. They all have a mission. If it's a state agency, it's a legislated mission. The Washington Department of Fisheries is told to do certain things. If it's an industry-funded organization, industry is paying into that for some reason. If it's the Fish and Wildlife Service, they've got a mandate. At a minimum there'll be some things that are in-bounds and out-of-bounds in terms of topics or species or something like that." Accepted methods and topics in any scientific field change through time, shifting the boundaries, as do politically oriented emphases.

Researchers at Canada's Pacific Biological Station experience the shifting political winds more directly than do most Americans. Although they do not have to apply externally for funding, they are at the behest of, as some there see it, political whim. Neil Self, who had recently retired when we spoke, told me that in his career

> the priority areas were shifting all the time. If you're going to do science, scientists, I think, should be put into types of work that they're interested in and they should be given time to accomplish something. To me that's the point. I think that isn't in conflict with sort of steering them into things that are relevant. But once you find something that's relevant, the issue is then to leave them there until they've gone somewhere. My perception is that we're not doing that now. I'm frankly concerned about what's happening to research because I think it's being affected too much by the weathervane, harum-scarum problems of management and too much by the weathervanes of politics.

Salmon biologists' research agendas are deeply affected by alterations in political inertia and momentum. They may be unable to pursue lines of research that are perceived as personally rewarding and professionally necessary because of the controlling influence of political and economic interests.

The discipline's historical yet sometimes tenuous ties to those interests almost have become its exclusive determinants. Salmon are re-created as political and economic pawns, not organisms that are perhaps important in these ways but that also are valued spiritually, for example, or in their own right, as they once were, and still are, to Native Americans in the region.[25] This has a profound effect on biologists' constructions of salmon. Salmon are seen by scientists through a sanctioned filter, and alternative viewpoints on an issue may be difficult to investigate. The salmon that is known to researchers may not be complete in a scientific sense, but the requirements of the economic and governmental institutions directing that research are not nearly so ambitious.

Expediency versus Knowledge

Because so many salmon populations in Oregon, Washington, and Idaho are facing extinction, the questions of most interest to BPA, the U.S. Army Corps of Engineers (which constructed most of the dams on the lower Snake River and the lower Columbia River that salmon encounter on their journeys to and from the ocean), and other governmental agencies in the United States and Canada have to do with near-term management issues. These are concerns such as the extent of predation by certain organisms on young fish or the number of adults that can safely be caught while guaranteeing that enough reach their spawning grounds or home hatcheries.

Such topics contrast with issues that many salmon biologists feel are most important, especially "life history" questions, some of which may only be effectively researched over generations of fish. These include salmon migration patterns, mate selection, juvenile behavior in streams, and differences in reproductive abilities between populations of salmon. Often they have little or nothing to do directly with the short-term economic or social benefits to be had from salmon, but in the long run biologists insist they may be essential if dwindling salmon runs are to be recovered.

This distinction between near- or short-term issues and long-term concerns closely corresponds to a rift in salmon biology between those whose work is "applied" and those whose work is considered "basic." Many biologists reject it, but the applied/basic parallel appears to have some validity, especially when considered historically, and it is appropriate to note here because of the increasing emphasis among funding agencies on short-term, applied research. One Canadian biologist said, "So there's constantly this budget tradeoff. To meet the demands within the Department of Fisheries, we have to do a certain amount of applied, short-term work with managers and for stock assessment. On the other hand, the science branch wants to be longer-term, and their outlook is a broader scale in the type of thing that we look to study. So there's always been this tradeoff developed." Scientists often find themselves frustrated, caught as they are between what scientifically "needs" to be done and what politically and economically "must" be done. They realize that budgets tend to be larger for applied work, yet they have been socialized to see basic science as the most thought-provoking.

An example of long-term research comes from salmon biologists' participation in interdisciplinary ecosystemic studies. For instance, the International Biological Program (IBP), established by Congress in the 1960s at the instigation of some of the nation's leading scientists, attempted to characterize and study the major biomes—distinct habitats—in the United States and Canada.[26] Little about the IBP was of short-term interest or emphasis, and ecologically minded salmon biologists—those of the "basic science" ilk—enthusiastically pursued IBP research.

Today, shrinking government budgets rule out not only geographically and temporally broad-based, "holistic" efforts like the IBP, but many area-specific and even species-specific ecological studies as well. In Canada, especially, the lack of funding for long-term, basic research and the pressure for quick and substantial economic returns from the nation's dwindling salmon runs has brought about a clash of outlook between basic and applied scientists. One Canadian researcher summed up the rift:

> The conclusion is that in many areas of our particular culture now, I think the perspective is that quite a lot of research is discretionary, that we're under incredible economic constraints, and that to the extent that you can identify research as a discretionary activity, we can ill afford it at the present time. So I think that's one of the things propelling the withdrawal of resources from it. There's an enormous, I mean an overwhelming, emphasis on applied research first and foremost. That's certainly apparent in academia, where indi-

> viduals would have paid allegiance to the notion, to the philosophy, of just doing research for the sake of research. But you can't gain grant support that way any longer. Certainly in government labs and central agencies, that sort of thing, any notion that you would do undisturbed research at some distance from an actual pressing question of the day that deals with resource management, any notion that you would do that with that remoteness to it has long vanished. There's just no room for it.

In Canadian government laboratories, basic research is a thing of the past, squeezed out by a corporate-state emphasis on present-day problems and immediate returns on research investments.

The outlook is similar in the United States. In both nations, applied studies are perceived as not only economically important, but politically so as well. Another Canadian, biostatistician Gary Hite, put it best when discussing researchers' applied orientation: "I think it helps you to get funds: show us something that's management-relevant. It's easier to show why you should get money to do it. It's easier to justify grant proposals or a hiring proposal. The fact that money is now out there for comanagement [Canadian-native peoples] stuff is what drives it, because they're not interested in your paper. They want to manage the fish: how can you increase the catch or what we should do to rebuild stocks." Similarly, while discussing the changes in the field just in the decade or so that he has been active, Tom Papas, a federal biologist in the United States, said, "Pork-barrel politics and 'me first' type things are still as rampant as ever. In terms of the research arena itself, one thing that is disappointing to me: things like life history aren't new, not flashy, but I think it's compellingly interesting. I think it's some of the most interesting stuff that we can look at. But that's not the way most people think about it because it's not real flashy stuff."

This distinction, as biologists draw it, is between economic and political expediency, on one hand, and knowledge, on the other. "Undisturbed research," life history studies, and the like are characterized by some as essential to the knowledge-creating mission of salmon biology. Yet that mission has always been rather schizophrenic. The "flashy" aspects of research are the ones that address social problems of immediate concern, such as catching adequate quantities of salmon or protecting depleted salmon populations, and they have always been part of salmon biology's reason for being. In contrast, basic research yields fundamental insights that may have a more lasting importance. Today, however, flash takes precedence over fundamentals.

Not all researchers have been thwarted in their efforts to obtain funding for long-term or life history projects. Paul McGuire, a salmon biologist who works for a state agency whose primary emphasis is on forests, not fish, said he has little difficulty finding funding for the life history research that interests him. His agency does not fund all of his research, making it imperative that he seek "soft money," such as grants from BPA:

> The funding is never everything you'd like it to be in total. But I look back over my career and research funding has not been a major stumbling block for me because I've always been able to sell good ideas and get the funding for them. I've mainly focused on the basic ecology of salmon, something that there hasn't been a lot of work in. There's been a lot of misunderstanding of the salmon, and a lot of people over the years have asked for different activities on the land on behalf of salmon. But we have a relatively poor understanding of the salmon. So I focused on understanding the salmon better, and it paid off well for me because it opened a lot of doors. It has received funding, and it has helped me to be more on-target as to what are the true needs of the salmon and where to focus my attentions. I've noticed over the years that many of the things that I thought of and put focus on later on "came into vogue," you might say.

The variation in responses regarding basic versus applied research—from those who are unequivocal in arguing for the importance of research that is of immediate benefit, to those who assert that basic research is being overlooked and is virtually unfundable, to those who have been successful in garnering funds where none were thought to be available—may be accounted for by several factors.

One is the *organizational location of the researcher*. Those in agencies with "missions" to manage salmon emphasized the pressures on biologists for immediate answers to problems; those like McGuire whose roles were secondary to an agency's goals, or who worked in academia, for the most part did not feel the same pressures to produce quick answers.

Sources of funding affect the emphasis on basic or applied research, too. The Canadian Department of Fisheries and Oceans biologists receive all of their financial support from the department itself, and the available funding for research is rapidly decreasing while the demand to catch more fish is increasing. This leads to an emphasis on applied studies. In the United States researchers told me that the major national funding sources, such as the National Science Foundation (NSF), make the application procedure laborious, and there may be a considerable delay between applying for an NSF grant and final word on whether it is funded. For these reasons, many researchers turn to the BPA and the Army Corps of Engineers.

The Corps and BPA may grant funds more promptly, but they also want relatively quick answers to immediate problems on the Columbia and Snake Rivers; despite there having been very little basic biological information (or "baseline data") gathered about the salmon in those areas, the precipitous decline in salmon numbers has led to a crisis-oriented approach by funding agencies. Now simply is not the time for long-term studies, say many researchers; unless the numbers of wild salmon adults returning to the region are stabilized, there will not be any fish left to research. In these days when salmon biology is enmeshed in the conflicting, simultaneous realities of local economic issues, global economic competition, endangered species, recreational anglers' concerns, and aboriginal rights, to name but a few, expediency is more important than knowledge.

Finally, a rough *demographic trend* is evident in the basic-versus-applied debate. The trend is demographic in two ways: both age and geographic location appeared to be important. For the most part, older biologists emphasized basic, life-history research, perhaps because they were trained when life history research was in vogue. Those younger than fifty were split in their outlook, and the first two factors above play a stronger role in explaining their perspective. Geographically, biologists farther inland—more distant from a fishing port—were more likely to emphasize basic research over applied approaches.

The Professional Politics of Funding

Having outlined the macrolevel realities facing salmon biologists in their search for funding, I now turn to a more microlevel emphasis. A favorite saying of the late Tip O'Neill that has become a cliché among politicians is that "all politics is local." Funding for salmon biology is local politics with a vengeance. At the macro level politicians, agencies, and corporations identify issues of particularly pressing concern; at the micro level, funds for particular research projects that address those issues are actually distributed. Many of the funding decisions in the United States occur in an almost face-to-face manner, as opposed to more traditional peer review of funding proposals. In the latter, a proposed research agenda is submitted to a potential funder, which distributes it to several scientists in the same field. They evaluate the proposal, and their identity is not known to the person seeking the grant.

This approach, long judged fair and ethical, often is not followed by funders of salmon research. Bob Nicks, an Oregon researcher, said that funding decisions are highly political, part of a "good old boys" network that primarily benefits those who are well-connected. He said funding was "politically motivated" and went on to explain,

> Since there's a finite amount of money, it's very competitive. It's food on the table, kids going to college, that sort of thing, and sometimes you don't get money. Fortunately, my salary's not involved, but there are a lot of people whose salaries are tied up in that. Well, it turns out that, particularly in large systems like the Columbia, you have so many agencies where there are biologists that also need this money. So you have biologists working for the agencies that are making decisions about the funding of those proposals. The people that make the decisions about which proposals get funded will make sure that one from each of the agencies that have people that sit on those committees get funded rather than which proposals have the most merit. In fact, there are people who sleep at these meetings and they'll wake up to vote for their stuff and then they'll go back to sleep. It's that kind of politics.

Hearing this, I was incredulous. I was astounded at the idea that "pork-barrel politics," Tom Papas's term for it, was at work in the scientific community *among scientists* (not among politicians seeking to fund particular studies to appeal to constituents) and especially within a group with, simultaneously, so much money available to it but also so much competition for the funds. Fairness did not seem to be a pressing concern.

Other researchers confirmed the veracity of Nicks's comment. For instance, in speaking of the Bonneville Power Administration's budget for salmon research, David Perkins was only slightly more subtle about grant-making procedures when he said, "It tends to be a somewhat conservative process because you've got a variety of interests literally as well as figuratively around the table talking about, 'Is this something that we would want to do?' So my sense is that it has tended to be a somewhat conservative program. It's difficult to get in, but once you're in, it's relatively easy to stay in." Those individuals and organizations with proven track records or who have a seat at the table are "in." And this in-group is the envy of many in the field.

The key point here is that salmon biology is controlled not only by external, macrolevel forces such as governmental and economic entities, but by microlevel social factors *internal* to the fisheries research community as well, including interagency politics and the elites that supervise many funding decisions. Belonging to an effective social network increases a researcher's chances of being funded, an observation that flies in the face of the "objectivity" that is cherished in the positivistic scientific paradigm. Moreover, this subjective approach to decision making sheds further light on the qualitative aspects of research. To the extent that funding decisions are based largely or entirely upon organizational affiliation, the questions and concerns that those organizations wish to investigate are the ones that will receive the most attention. Other potentially illuminating biological questions and means of studying them may be ignored. The salmon that results is a politically sanctioned construction, not only at the macrostructural level of national or regional/state policy, but also at the microstructural level as well, where professional, personal networks come into play.

"Bootlegging" Research

Salmon biologists are not helpless in the face of stiff competition for research funding. I was told, for example, that the successful biologists or those with effective networks may be able to position themselves to work on emerging, "hot," fundable lines of research. Biologists also possess other means for securing funding as best they can. One way is by "bootlegging" the initial, unfunded research for a project onto an already-funded research endeavor.

Generally, federal, state, provincial, and private organizations award grants for *specific* research projects. Bootlegging is the use of funds given for one study for a different but related purpose. David Perkins, an American with research experience on both sides of the border, explained, "In fact, I've never had explicit funding for most of the interesting things that I've done. Most of the really intriguing things have been kind of bootlegged on the side. Only a few of the important things have had that explicit funding to do what was proposed." He later added, "When I was at PBS [Canada's Pacific Biological Station], some of our neat stuff was done on the side. You went ahead and did it. As a post-doc I had fairly free rein, and most of the stuff I did really didn't cost anything. Pretty cheap." Bootlegging appears to be a common practice and a common term in the profession on both sides of the border. Another biologist, the geneticist Alex Stand, elaborated on the importance of bootlegging to researchers:

> I think, again, we just have to prioritize the projects. If there is something that is just overwhelmingly exciting in our minds, we'll try to find a way to do it. In reality, because you have a grant on one topic doesn't mean that you can't, on the side, do small experiments on another topic. Most of the researchers that I know approach things that way. They spend a fraction of their time exploring interesting new directions while they are doing the mainline projects. That's how you get into new projects. In reality, in order to get money for a new project these days, you can't do it just solely based on a good idea. You have to have some preliminary results to sell the project when you write the proposal. And, of course, you could never get into a new area, therefore, unless you bootlegged a little bit in how you spend your effort.

Bootlegging, then, may be the only means of garnering the initial data necessary to support a grant proposal. Nor is bootlegging a new or uncommon phenomenon. Referring to the research climate in his initial years in the profession, a salmon hatchery biologist with more than twenty years of research experience told me, "I think our problem was that we were not doing good research in the early times. A lot of the research, there wasn't a lot of dollars in it. So you usually had work kind of tacked onto something: 'Let's mark a few of these fish and see what happens.' Well, you end up with a sample of one for one year and then you compare it with the next year!" Bootlegging provided a data-based start for new studies, however weak the logical or statistical foundation.

A university researcher with substantial experience in the field alluded to bootlegging more subtly. Commenting on his work with a species of fish that has coexisted with salmon for thousands of years but that is considered a nuisance in another area of the country, he said, "The hidden agenda that I have is we're interested in preserving [the coexisting species] out here, which are in trouble now. We're working with Pacific species, and there's virtually nothing known about their biology and their culture, so it's worth getting into that with money from the [studies elsewhere in the United States], where [the coexisting fish] are a problem." (I have omitted the details to ensure the researcher's anonymity.) In this case, research funds for a study in a distant geographic locale were used as seed money for a local study relevant to Pacific salmon.

While bootlegging is commonplace in many fields, it was fascinating to hear it so openly discussed, especially in light of the competition for grants. But that is precisely why bootlegging takes place. It is an understandable unintended consequence of the closed-market orientation that characterizes salmon biology funding. It simultaneously reflects researchers' anxieties about maintaining the currency of their projects and their struggles to regain some control over their endeavors.

As for the potentially inflammatory admission that researchers bootleg projects, bootlegging does not appear to be illegal. Regardless of its legality, however, what is of sociological interest is that researchers do not consider bootlegging improper in a normative sense. That is, bootlegging is an accepted part of what salmon biologists do, plain and simple, legal or not. Given that the issue was raised by four biologists working in three different states, it seems justifiable to say that the practice is widespread and even taken for granted: expected, unchallenged, and necessary for professional success and personal fulfillment. Bootlegging is a norm among salmon biologists, and it may be seen as a form of resistance in the face of the " 'embedded power' " of the governmental and economic forces constraining salmon biology.[27] If this is the case, it may also be that funding organizations are aware of the practice and may sanction it, since they have not moved to stop it.

However, bootlegging has its limits. Even it is controlled. In particular, funds from organizations with broad mandates are available for bootlegging, but research sponsored by fisheries-oriented agencies is not. David Perkins observed, "Problem is with agencies, they want to know in advance what they're going to get out of it. If you start wandering around and following interesting leads, they're not particularly pleased with that." Similarly, Alex Stand commented, "I think that's something that the agencies all realize, and agencies like the NSF or NIH [the National Institutes of Health] are pretty willing to understand that. I think with an agency like the Bonneville Power Administration, you try a little harder to strictly follow the guidelines because that isn't what they're about. They aren't about supporting basic research. They're about getting a particular task done. So I think you try to weigh things accordingly." Researchers must be careful whose money they take liberties with, and they learn these parameters and other nuances of bootlegging as part of their professional socialization as graduate students.

In contrast, in Canada bootlegging may take on a deliberate, acknowledged, and even rewarded form. George Williams, one of Canada's most revered salmon biologists, told me,

> It's so liberal a system that you can write up, "I want to do research on houseflies and their population dynamics, where do they all go in the winter time?" You want $30,000 to do this. The committee will say, "That guy really knows his stuff, so we'll give him his $30,000." Then, after you receive the money you say, "You know, a lot better question would have been to ask about the bumblebees. What do they do in the winter time? So I'm going to do research on bumblebees instead." You don't even have to write and tell them. You're given the money on the strength of your making a case. If, at the end of the grant, you've published a half dozen papers on the bumblebees and what they do in the wintertime, people will say, "Isn't that great. He said he was going to work on houseflies and he had the guts to turn around and work on bumblebees because he saw that it was a better opportunity."

When Canadian researchers have a brainstorm, they are expected to pursue it. This reflects the contention made by several biologists that the Canadian funding approach for university researchers (though not for their counterparts in government) is substantially more liberal and flexible than in the United States. The Canadian government vests researchers with responsibility and rewards them for contributing to scientific knowledge. Compared to the expectations of agencies in the United States, the Canadians, while still controlling researchers, allow a substantial amount of self-direction to researchers who wish to follow a "better opportunity."

In the United States, savvy, well-socialized salmon biologists know that bootlegging is an option pursued by many researchers and that it is not an accepted practice with some funding sources, primarily agencies that specialize in fisheries issues. Alex Stand's connection of bootlegging with "basic research" and with agencies that support such research is especially telling. It indicates that, although applied research in this field garners the bulk of the funding, certain agencies are seen as supporting basic research, and researchers are aware that with this support goes the tacit acceptance of bootlegging. I suspect that another social phenomenon is operating here, one that sociologists have not attended to extensively, perhaps because of their historical interest in issues of control. *Freedom* best characterizes researchers' creative approaches to pursuing promising leads. In this instance, freedom is an attempt by biologists to resist control by others, and it leads to strikingly different constructions of salmon. I develop this concept further in Chapter Six.

CONCLUSION: BIOLOGISTS AS BARTLEBY

What can we say about biologists' social constructions of salmon at this point? Several conclusions emerge. In the guise of professional expectations, control was part of salmon biologists' realities from the establishment of the discipline. Their world was shaped by political and economic factors that could not be separated. That is, governmental assistance for the commercial and recreational fisheries community was an early priority for those in the discipline, and public policy and private gain have been intertwined ever since. This reflects themes identified by Max Weber. Weber's interpretation of the emergence of capitalism in the West was that it "is characterized, above all else, by the absolute, methodical control of every activity of daily life."[28] Science embodies this control like no other social institution, and Weber identified control through science as fundamental to the rationalization of Western life. When the object of interest was biological, the control imperative was only heightened, because biological entities can grow, swim away, and otherwise "resist" control.

This political-economic presence has had far-reaching implications for salmon biology and its constructions of salmon. Until recently biologists have been able to entertain an expansive view of what constitutes *fisheries research,* and this has allowed them wide latitude in the meanings they have given to salmon. In the "old days" that came to an end in the 1980s, a biologist could study the details of salmon "life history" such as mate selection and juvenile salmon behavior simply because the research filled a gap in the knowledge about the fish. Salmon were a fascinating biological entity worthy of study in their own right. If that new knowledge could be used to fulfill economic or policy goals like improving runs or catches, that was all well and good, but it was not the sole reason for conducting research. The discipline still was the primary force directing researchers' constructions of the salmon.

But today salmon biology's historical yet sometimes tenuous ties to economics and government almost have become its exclusive determinants. With that has come a narrowing construction of salmon. Increasingly, salmon biology has become embedded in the structural processes that gave rise to it. Of utmost importance was and is the funding of research. Money flowed to salmon biology because of the belief by governmental and private entities that science held the solution to creating more salmon, a belief that persists today. Yet, increasingly, research is characterized by short-term issues, such as the expected returns of fish to a given hatchery or river, rather than "basic" research into salmon life history, which some biologists argue is the foundation for the more applied approaches and for lasting salmon populations.

The movement of funding toward near-term issues and away from ecological studies reflects diverse structural and biological changes, such as the ongoing fiscal crises experienced by North American governments, the decline in the number of fish outside of Alaskan waters, and the corresponding identification of threatened and endangered wild runs of salmon.[29] The result is that funding decisions, including those overseen by biologists, are increasingly weighted toward research that promises near-term solutions to problems identified by policy makers.

Like Herman Melville's Bartleby, who is a metaphor for the writer condemned by convention to copy the styles and ideas acceptable to publishers, biologists find much of their research creativity eliminated by the vicissitudes of institutional forces beyond their control.

Bootlegging demonstrates that salmon biologists, regardless of the other nuances of their constructions of salmon, are driven by professional, political, and economic pressures to create new knowledge. To develop these new knowledges, they resort to creative uses of existing funding to launch new areas of investigation. Sociologically, what is exciting about bootlegging is that it is a window on how salmon are reconstructed, given new meaning, by biologists. In bootlegging, as in all funding efforts, we find these scientists compelled by status concerns—including the need to retain their jobs and to improve their standing in the biological community—to develop new insights about salmon.

As Berger and Luckmann suggest, with this new knowledge come new constructions. Perhaps the knowledge produced indicates that salmon navigate to their natal streams using the earth's magnetic field rather than the stars, as once was believed. Or maybe the new knowledge from bootlegging indicates that steelhead trout ought to be included in the *Oncorhynchus,* Pacific salmon, genus and not with the trout in the *Salmo* genus.[30] The details of the knowledge are rather immaterial. The importance of it for us is that, whether through bootlegging or not, new knowledge produces new social constructions. Once the biological community places its stamp of approval on research, as it does whenever research findings are published through the peer review process, salmon become new creatures; suddenly the imprimatur of science dictates that they are compelled by magnetism to their homes or that they are related in ways no one had previously considered. Whatever the case, salmon mean something different. And because knowledge and meaning shape our perception of reality, each new insight created by research changes the salmon world as scientists know it a tiny bit.

3 Biologists in the Driver's Seat

IT WAS ESPECIALLY exciting for me to speak with Klaus Huberman. Not only did he sit for one of my first interviews, but his research on salmon behavior is among the most respected in the discipline. Gregarious and silver-haired, he seemed like a father figure to the younger biologists who interrupted us with questions several times during our discussion. Huberman was a forceful advocate for a biology-first approach to research, upset as he clearly was with current trends that curb researcher freedom and give a dominant place to politics, economics, and, increasingly, engineering. The latter, especially, troubled him. "This is the thing we're fighting very hard," he said.

> The engineers are too quick to propose engineering solutions. And if you look back in history, you find out that most of them were utter failures, because our understanding of how an ecosystem works and what a habitat is and how animals behave and what the fluctuations are and the balances is really very poor. So therefore we may think that we have an engineering solution that might work for a few years and then, bango!
>
> .... So our advice is, don't fiddle around with it so long as you really don't know how all this fits together. The engineering solutions, we've heard them all, and in the long run they're all failures because they don't deal with all of the different aspects.

For all of his earnest sentiments, however, there is an irony in Huberman's language. Although he finds himself "fighting" an engineering perspective, the *system* concept in the commonplace term "ecosystem" has been strongly influenced by engineers. Indeed, in an effort to bring conceptual order and control to their understanding of salmon, biologists rely a great deal on a vocabulary that they share with engineers. Despite the irony, there is some logic in this, for no other profession has more success in transforming the mysterious and uncertain into the known and predictable than engineering. Transformations of the same sort are seen by biologists as their highest and best calling.

CONTROL OF SALMON BY SALMON BIOLOGISTS

It is impossible to overestimate the importance of controlling salmon for the successful pursuit of salmon biology. Essential to this norm of control is the creation and demonstration of an *engineering model of salmon*. This systems view of salmon and the salmon's environment is developed by biologists in laboratories, field studies, and in salmon hatcheries. Ultimately the systems perspective depends upon the successful quantification of the fish and their behavior, enabling salmon to be portrayed in an abstract manner such that anyone with the "proper training" can comprehend the scientific construction of the salmon: what salmon really mean, according to biologists.

ENGINEERING SALMON

Historically, the engineering ideal that salmon biologists adopted may have come about because engineers possess immense power and control relative both to salmon and to those who work with salmon, as Huberman implied. Engineers are so successful at what they do that they have been portrayed as enemies of the salmon. It was engineers who designed polluting industrial factories, elaborate fishing gear (including products as diverse as net materials and sonar), and unyielding concrete dams. Indirectly, they often made life more difficult for the salmon. Arguably, engineers more than any other social group have been responsible for extirpating salmon from their home streams. In another irony, many biologists would never have had jobs were it not for those engineered impacts.

Oddly, this is precisely why engineers are the ideal. They, more than any other social group, enable humans to exert control over Nature. Biologists recognized the power of the engineering model long ago. Philip J. Pauly wrote that in the last century "biologists began to think of themselves and their work within the framework of engineering. They argued that the fundamental purpose of their science ought to be the control of organisms."[1] It would not be correct to say that the biologists of today consciously study engineering or even that most are aware of the omnipresence of the engineering ideal in their work. Nevertheless, what greater testament to the power of an idea is there than for it to become a taken-for-granted, part of the background reality of a group that so often disparages that very idea?

Engineering possesses all of the characteristics that many biologists identify as penultimate goals for their science. First, engineers have effective conceptual models, namely, theories. Second, these theories posit the existence of systems; actually, they *create* systems, for systems do not exist apart from our calling them forth. Third, those systems *work.* That is, the "behavior" of those systems is predictable—what happens in engineered systems is "known," is certain to a high degree. Both the predictability and the degree of precision exist because these systems have been quantified to an extreme. Prediction is the teleological end product of the union of systems with quantification, foretold in the use of numbers to describe everything in a system in the first place.

Biologists may not acknowledge it even to themselves, but they look at engineers with envy. From the earliest days of their training at fisheries schools, biologists are subtly inculcated with an engineering-like norm of control of the system as the desired condition, whether that system is an ecosystem or a salmon's nervous system. Yet they live with the anxieties of poorly theorized, seemingly uncontrollable (and therefore nonexistent, in a strict engineering sense) systems that defy quantification at levels of specification that even remotely approach those of engineers. One biologist commented that his colleagues work "with the knowledge that, unlike engineering where your R^2s will be real high, they're typically lower to begin with in biology." That is, in engineering virtually every potential unknown *is* known and can be enumerated; such is not the case in biology.

Another scientist observed, "We know a lot. We don't know things in black and white as well as we want to know. Engineers get frustrated with biologists because they can't tell them exactly what a fish is going to do in the same way that an engineer can measure what it's going to take to bend a piece of steel. And that's just because of the inherent variability of biological creatures. You can tell central tendencies of what things are going to do. But they're going to react to changes in the environment, and we're not always sure how they're going to react. As we learn more and more, we get better and better at predicting."

Engineers succeed because their predictions are so exact, and this is possible because their definitions of systems bring conceptual closure to nearly all variables, all of the identifiable things that are of interest to them. This means that every possible variable and every possible attribute of the relevant variables is taken into account when, for example, a dam is constructed. This closure actually describes a *closure of meaning*. When a dam is complete and operates as designed, for example, its meaning as a successful engineering achievement is affirmed, and all further debate about its effectiveness ends. It is left to engineers or scientists to identify that moment of closure when all that needs to be known is known about a problem, such as "will the dam hold?" or "how do salmon find their way to their spawning grounds?"

But also of interest is the *opening* of previously closed social groups like engineers or salmon biologists, or the opening—the challenging and questioning—of the decisions made by those groups. Typically, these groups and their decision-making processes are obscured from public view. But they often are available for manipulation by other groups, especially those with substantial amounts of control and power.

An example of the opening of engineers' "black boxes,"[2] and therefore an opening of a previously closed system, is in order. Such an event occurred in 1992 when a "test drawdown" was conducted of the reservoirs behind two Snake River dams. Drawdowns—the deliberate, rapid, and extreme lowering of water levels—are advocated by some as the easiest, surest way to help dwindling wild salmon runs recover. The reservoirs are long, thin lakes hemmed-in by basalt canyons. If their levels are lowered, the Snake River's channel will narrow and the water will flow faster. Young salmon swimming toward the ocean will survive in greater numbers, the theory goes, because predators like squawfish and gulls will be less likely to kill them, thanks to the increased speed of the water.

The 1992 drawdown opened the engineers' world, allowing in uncertainty by introducing new, "unknown" variables (things without data) to their equations. Previously all variables had been known, meaning that the "behavior" of the dams, the electricity-producing turbines in their bowels, and the like was predictable. When the dams were designed and constructed, engineers assumed that a reservoir's depth would never fall beneath a certain level. This is an example of a "scope condition," something that is specified in order that a theory be predictive. (The statement, "The temperature must be zero degrees Celsius or lower" is a scope statement for the theory that states, "Water freezes.") When engineers were forced by social groups' demands to violate the scope conditions for reservoir depth, the engineers felt the frustration of uncertainty that salmon biologists routinely experience. One biologist was quietly gleeful as he mused about the drawdown: "But it's interesting to listen to some of these engineers talk about whether the turbines down at the dam can operate at reduced head [lower reservoir water levels]. They're not at all sure. . . . They kept running them throughout this test, but they're not at all sure that they can do it year in and year out." Nature was peering into their black-box system assumptions.

Systems

As mentioned, salmon biology shares its key metaphor with engineering: the *system*. Biologists speak of river systems in which the salmon live; they speak of hatchery systems, which may either refer to the hundreds of salmon hatchery sites spread throughout the region or the "systemlike" operations of a particular hatchery; and they often use the word *ecosystem* to describe a salmon's environment.[3] John Law writes that the systems metaphor has advantages and disadvantages; systems include and exclude, and in so doing they bring a degree of closure to a phenomenon that is never truly closed: "The metaphor stresses heterogeneity and interrelatedness, but it also tends to direct attention away from the *struggles* that shape a network of heterogeneous and mutually sustaining elements. System builders try to associate elements in what they hope will be a durable array. They try to dissociate hostile systems and reassemble their components in a manner that contributes to what is being built."[4]

Systems are conceptual, and they are created by engineers and scientists. The problem with seeing salmon as part of a system is that, as living biological creatures, they resist this network-shaping. Salmon almost seem to struggle against systemization just as they fight to swim upstream against river currents from the ocean. Salmon are not hard concrete or cold steel or electrified computer chips. Those things "behave" for engineers the same way every time, though "behave" is a crude metaphor. They function. Biological entities like salmon really do behave. To salmon biologists' chagrin, often they behave rather "badly." They refuse to obey engineering-like laws. They act unpredictably. Biological systems are intensely uncertain, and because of this they fall well short of the engineering ideal.

There are 450 dams in the Columbia and Snake river system, and on the Columbia and Snake Rivers themselves there are two dozen of these huge water-holding and power-producing dams (enough to provide far more electrical power and irrigation water than the citizens of the region actually use, and this makes both very inexpensive). This already-engineered status leads some biologists to argue that the only solution to reversing the system's dwindling salmon runs is more engineering. Jim Winston put the situation—and the solution—bluntly:

> I think they should have simply looked at the river and said, "We've developed it to the point where it's an industrial river. We're not going to ask you to change that use of the river. We're just going to bypass it. We're going to move the smolts [juvenile, oceangoing salmon] into a canal and move them downstream." We'd have no predator loss, have no temperature loss, no supersaturation problem, no spillage loss with the dams, and they'll have a new route that they'll go down. It'll cost about $500 million, but that's a small price compared to the adjustments and the loss of income those adjustments are going to cost the area.
>
> That's contrary, conceptually, to the naturalist who says, "Don't give up the river. That's where the fish belong. Do whatever you have to, but don't give up the river." I don't think there's room for that in this day. I think the river is a happenstance of geology. It could have been one hundred little rivers. It happened to be one big river. Those populations are happenstances of the circumstances that allow them to colonize new environments. Had we had a fifty-foot Celilo Falls there'd have been no upstream populations. So let's not look at preserving nature for nature's sake. Let's look at how we can enhance our resources by appropriate alterations to make use of them. You probably have a whole range of opinions that that sets

> into. It doesn't reduce the importance of that population. It doesn't reduce the importance of managing the populations for their biology and the way that they fit into the system. It means making sure that our alteration of the environment can be in some way neutralized and still have the benefit of those developments.

In the region's highly engineered Environment, the only way to save salmon is to engineer the system even more.

Other biologists reel at such suggestions. As Klaus Huberman commented at the beginning of this chapter, engineering solutions to biological problems are no solutions at all: "And if you look back in history, you find out that most of them were utter failures, because our understanding of how an ecosystem works and what a habitat is and how animals behave and what the fluctuations are and the balances is really very poor." Biologists like Huberman portray themselves as living in another world from engineers. Even some hatchery biologists, who work in a highly systematized, engineered environment, are skeptical of engineering solutions. As one of them observed, "But how are they going to fix it? What do you do? An engineer is not a biologist and a biologist is not an engineer. How do you put the two together?" Ignored by all of these biologists is that the engineers already have won the day whenever systems thinking predominates the discussion.

The systems perspective does deserve its due from biologists, who use it constantly. It *is* illuminating. It demonstrates fascinating, thought-provoking, and often reliable interconnections between the chosen phenomena, even if, as John Law noted, much is omitted in the process. However, systems may be most powerful as speculative and heuristic—that is, not when viewed as the real thing but instead as educational tools or as means to improve thinking and prompt new insights. An especially poignant example of this was a comment made by a biologist who worked on the Snake River drawdown project noted above. She mentioned that crayfish failed to move with the water as the reservoirs behind dams were lowered. The crayfish burrowed into mud but eventually died when the mud dried out. "People would argue, 'Well, so what? Who cares about crayfish?' " she said.

> Well, two major points on crayfish. Number one is, they are a favorite food of the squawfish and the smallmouth bass [both of which feed

> on smolts when crayfish are not available]. What the ratio of crayfish to smolts are when the smolts are going through [the reservoirs], I don't think we really know. But [one researcher] has observed that it appears that the consumption on smolts was higher this year. If your favorite food supply disappears, you're going to eat more of something that is abundant, maybe less preferred but easy to get. What's the net effect? We don't know. And we can't study it now because we don't know what it was to know what kind of effect we had.

Systems are about connections. Yet, as this example demonstrates, salmon biologists often are unaware of the existence or extent of connections between the organisms and other phenomena within the systems that they have identified. This demonstrates the heuristic aspect of the systems metaphor. Systems are conceptual guides for salmon biologists far more than they are physical entities or theoretical equations.

And why not heuristics? After all, from the biologists' point of view, systems are almost always incomplete. Can they ever be anything other than teaching tools? Neil Self, who had been involved in a large-scale forest-salmon ecosystem study late in his career, said,

> We don't understand enough about the basic ecology of young salmon, say chinook or sockeye in streams, to be able to talk about them and make good management decisions about them in terms of land use planning. We're still behind the eight ball. The project I worked with, we had some eighteen or twenty years of information that focused primarily on chum salmon, coho salmon, and steelhead. We know a little bit about their biology and what the implications are of forest management practices in one kind of system. But we should really have the same kind of work in systems that behave very differently.

Biologists do not like uncertainty and guesswork any more than engineers do. Often, though, they are left with little certainty and much speculation. In this case, no money was available for long-term research in different systems. In others, salmon biologists are unable to control salmon and other aspects of the ecosystems in which the salmon are embedded. The best that those using the system concept can do in such circumstances is to educate one another, not attempt predictions.

Engineering is a Weberian "ideal type" for salmon biologists. Max Weber's method is best characterized by the use of ideal typical examples, theoretical extremes that are nonexistent or unrealizable in social practice yet are useful for comparison purposes. Social groups often adopt ideal types and reference groups themselves, and such is the case with salmon biology's longing looks toward engineering. In engineered systems everything has its place, and the same is true for biologists' ecosystems. However, salmon almost seem to resent, resist, this placement. Biologists want to construct them as (Natural) law-abiding and certain, but salmon behave in unexpected ways because so little is known about them. The engineering ideal directs that everything is known as a scope condition. Nothing is left to chance. But biologists cannot construct salmon as known, for the one thing that the biologists know for certain is that the salmon are not predictable. This is precisely because the systems in which salmon live are never totally closed.

LABORATORIES, FIELD RESEARCH, AND CONTROL

Laboratories offer biologists a low-cost, high-control refuge from the chaos and confusion of substantially open systems like rivers. A laboratory, according to Bruno Latour, is "the place where scientists *work*."[5] Laboratories are locales, though certainly not the only ones, of scientific behavior and activity (also especially important in salmon biology are field studies, which take place entirely outside of the laboratory proper, although perhaps nowhere is "outside of the lab" in Latour's definition). Much can be (and has been) said about the activities of scientists in their laboratories.[6] Little if any of it, however, has explicitly approached the role of control in laboratory settings.

Control is integral to what goes on there, and for most researchers control is *the* reason for working in a laboratory in the first place (even if a researcher brings fish into the laboratory merely to make viewing easier than it is in murky stream water, the conditions are nevertheless being controlled). Sam Freiberg, with considerable research experience inside and outside of the lab, told me "laboratory-type stuff" was "where we control all of the elements and situations where we bring fish in." Another biologist contrasted the assurance of laboratory work, with its highly controlled surroundings, with the uncertainty of field work, and at the same time gave credence to Latour's view of the omnipresent laboratory, saying,

> Good science to me in an applied context is using a scientific approach to the question. Can you pose it in such a way that you can actually test an hypothesis and try to learn from your experience? At the same time you want to build that into the real environment. If you're going to study a hatchery, there's no point in removing them and working in a very isolated laboratory. You might as well work in a large production hatchery. So you work in the real environment. If you're going to study a mixed stock management issue, then you've got to work in the environment. So a lot of times what this does is remove a lot of the control. So you've got to be quite a bit more creative in terms of how you'll design your experiment, what are your controls, what are you testing, that sort of thing.

"Creativity" becomes a central issue of concern in minimally controlled experiments, such as those undertaken in the field. In field work the controls often are statistical, but in laboratory settings control over the salmon is exerted physically.

The first step in the process of controlling salmon and their surroundings in the laboratory is the removal of fish from streams (this includes using fish raised for research whose ancestors were taken from streams). Once in the laboratory, the manipulation of the salmon and their surroundings may proceed in countless ways. Biologists told me about "stress tests" in which fish swam against onrushing water until they were exhausted; testing ultraviolet visual sensitivity using probes placed in the optic nerves of fish; examining in-stream behavior of numerous types, including spacial separation between differing species of salmon by placing them together in the same tank; and forcing salmon to spawn in laboratory tanks to test their ability to clean gravel of silt.

These examples indicate a number of things, including the vast range of questions being investigated by salmon biologists and the biologists' abilities (both conceptual and financial) to pursue those questions. But what is the value of such knowledge? That is, how *transferable* is laboratory work to the "real world?" The answer depends upon the biologist being asked, the experiment being performed, and the chosen measure of the concept "value." Ian Smyth, an experienced researcher at

Canada's PBS, responded to my query regarding the transferability of laboratory results to wild salmon populations, saying, "That's a tough question. A lot of the work that I do in the lab is behavioral, so that sort of question comes up all the time in terms of what you learn in the lab isn't transferrable. My feeling is it's more transferrable than we think. Because we're looking at the eco-genetic issues in many cases that I'm studying, if there is a true genetic basis to it, then you can find ways of measuring it. It might not be expressed in the same way, but I think if you find something that is consistent between replicates in stocks and this sort of thing, then it probably will be expressed in some way in the natural system."

Mark Johnson was more adamant: "I think there's quite a lot of realism in the work that I do. I think we're a pretty no-nonsense group. We cut to fundamental questions that have applied outcomes and we have a fairly high frequency interaction between what we're doing by way of synthesis [in the lab] and what we do in the field going back to look at what syntheses we've done hold up and whether they stand the test of repeatability in real population or real ecosystem performance, depending on whether we're doing the population or community or ecosystem group." Others are less certain. Neil Self said plainly, "My lab work mimicked some aspects of the wild circumstance very well. But it also specifically ruled out some things."

Still others alluded to the complexity of the Natural environment. Pam Mitchell, who works for a federal agency in the United States overseeing a range of field and laboratory studies, commented on the transferability question by remarking,

> There's a lot of question with that. They give you some good information, but it's always questionable. Like the dissolved gas measurements, for example. [Dissolved gasses are produced by air mixing with water, in this case when it flows over dams, and high dissolved gas levels can kill salmon.] A lot of our tests early on were with fish in fairly deep tanks that gave the fish a range of depths and we could see what the mortality was at different depths. But how does that translate to reality? You might get a measure of mortality in that situation. But when it's coupled with how it affects the fish's ability to respond to predators or maintain a position in a water column which isn't safe and stable or whatever, we don't get answers to those questions.

This response implies not only that research may not be transferrable from the lab to the field, but that control can create its own problems when laboratory results are presumed to reflect nonlaboratory settings. Just as systems are only conceptual and "in reality" can never be completely closed off from external forces, so, too, is it the case that scientists' experiments—experiments in control as much as anything else—are limited by the very effort to control. Every theory specifies its scope conditions, and scope conditions are an admission of the limits of control. They tell us when an experiment will work. Under all other conditions the relevancy of the experiment's results are unknown. Under *those* conditions the same experiment risks failure. Control is incomplete.

QUANTIFICATION AND MODELING

Biologists, like engineers, quantify variables and include them in elaborate mathematical equations. Bruno Latour has argued that scientific literature can be defined "by its most obvious trait: the presence of numbers, geometrical figures, equations, mathematics, etc."[7] He noted the importance of equations in particular because he saw that they were used by scientists to "tie different things together and make them equivalent;" he went on to say that the "equations produced . . . tell us what is associated with what; they define the nature of the relation; finally, they often express a measure of the resistance of each association to disruption."[8] This definition of equation is strikingly similar to the definition of "system" as an association of durable elements. (The two concepts often are brought together, as when reference is made to "a system of equations.")

The process of quantification is concerned with developing and maintaining the basis for associations of phenomena. The chosen phenomena are, according to Latour and others, the result of social processes, not Natural ones: for instance, when closure is achieved and a mathematical model is deemed complete. Social structural processes play a role in this as well: are there limits to funding? (There always are.) Does the political or professional will remain to pursue the model as originally envisioned, or will certain variables have to be eliminated and assumptions added to the model to compensate for missing data? My fieldwork and discussions with modelers led me to believe that such pressures are always present. Society is in the equations along with Nature.

Quantification or allusions to quantification (as when biologists spoke of their "data," for example) occurred in all of my interviews. I was especially intrigued by the frequent references to "models:" quantified, computerized systems of equations, the output from which could be used for any number of purposes, including policy creation and justification. These comments were especially provocative because models are highly complex; the modelers with whom I eventually spoke left me with the strong sense that no form of equation building that can be done is as ambitious as that undertaken in modeling. Models' mathematical sophistication is substantial, and to generate the data used in them, large numbers of variables first must be identified. In turn, quantified observations must be gathered for those variables, usually over considerable amounts of time. Given the uncertainty and variability that scientists argue is inherent in "natural systems," the end result of this quantification process is inevitably less than perfect.

An illustration from my fieldwork will explain what I mean by this lack of exactness and by my statement that society is in the equations. In 1993 I attended a hearing in Lewiston, Idaho, sponsored by the Army Corps of Engineers. The Corps called the hearing to present its projections for future salmon runs in the Columbia-Snake river system and to receive the public's comments regarding the various "management options" that it presented. The projections for adult salmon returning to the Snake River varied with the potential changes in dam operations and other factors. A series of slides were shown, including several graphs presenting the results of the Corps's mathematical models. I was struck by the upward slope of the lines in many of the graphs, indicating constantly improving fish runs in future years. The number of returning salmon increased regardless of the changes in conditions. Moreover, the lack of ranges in the projections gave the impression of certainty. All results were single, straight lines rather than multiple lines or at least wide bands surrounding the single lines, which would have implied that a variety of outcomes were possible.

Afterwards, I asked the Corps biologist who presented the modeling results if there was an "R^2" or other statistic that allowed him to determine the amount of variance accounted for by his equations; basically, I was asking how complete the models were. Given the bold predictions in the slides, the data for the variables must have been of very good quality and the equations highly reliable. He answered, "Well, there are *multiple* multiple regressions in each of these formulas, so each one has an R^2." "What is the value of those R^2s?" I asked. He replied, "Oh, usually about point-four or point-five."

It was one of many moments in this project where preconceptions about biologists and biological systems were shattered. A "point-four or point-five" R^2 meant that the equations were woefully inadequate for policy-making purposes. They only "captured" 40 or 50 percent of all of the information that there was to be had about the region's salmon runs. The other 50 to 60 percent was left unexplained. So those neat, single line, almost always upward-sloping projections portrayed in the graphs—graphs that might be used to determine the future of salmon recovery in the Pacific Northwest of the United States—were unable to account for even half of the forces affecting the salmon. Of the one hundred or more persons at the meeting, no one else asked questions like the ones I posed afterward.

The public was meant to see a clear, unambiguous picture in those graphs. It was not meant to see society in that picture, only *nature*. Obviously, the Corps felt compelled to produce a convincing vision of the future of the salmon, and what better way than with colorful graphs backed up by the best science available? But the public was not told how the variables were selected for the equations that produced the projections upon which the graphs were based, and they were not told of the poor quality of the data supporting those variables. "Society" was present in this science in two senses. One was the Corps scientists' decision to include certain variables rather than others, a process that I was not privy to but that must have occurred. The second was the political decision to withhold the poor quality of the models from the public. That decision may have been made for any number of reasons, but certainly the public was not given the "complete picture" as the Corps scientists understood it. I left feeling that the presentation had been less science than show.

According to some biologists, the use of models as planning and policy-making tools by the Corps, and the modeling of expected future behavior as if it were certain, represents the wave of the future in biology. Ian Smyth took this position, saying, "I really don't see any other way. I've spent a lot of time sitting down and just trying to chat through these things, and it doesn't work. . . . I think we're going to spend a lot more time in an increasingly complex situation trying to resolve these multiple resource use management issues. I think that large-scale environmental modeling will become more important." He felt that modeling is the ultimate achievement in the march toward quantification, and thus scientific knowledge of the salmon. Scientists in numerous disciplines appear to agree; the journal *Environmental Modeling and Software* is in its second decade, indicating substantial interest in the area.

In Smyth's view, quantification in general and models in particular capture complexity—in effect, they extend control—by condensing systems into a manageable form (numbers and equations) so that those who are properly trained can determine natural resources policy. This is a form of "scientific management," not in the turn-of-the-century sense, when laborers were portrayed as cogs in industrial machines who could be made to work more efficiently and productively through the application of pseudoscientific principles, but in a new sense where science is identified as the arbiter of all social-Environment conflicts. Politicians invoke science when they speak of Nature. So do corporations. And even the most radical environmental groups have their own pet scientific paradigm.[9]

For salmon biologists there also are status issues involved with quantification and modeling. George Williams remarked that in mid-career he "realized that there are parallel streams in the literature, so you can look at what is being done in fisheries and look at what's being done in wildlife, all the same issues. Fisheries is running quite a bit behind in the degree of mathematical sophistication. Forestry, a similar stream of literature; forest entomology, for example, all the same equations." Williams became concerned enough about his discipline's lack of familiarity with cutting-edge quantitative methods that he took a sabbatical for the sole purpose of acquainting himself with the leading mathematical techniques. He soon became a strong proponent of modeling.

This was the 1960s and early 1970s. Large-scale modeling projects were difficult undertakings, not least because of the poor computing capabilities available to researchers compared to the powerful hardware and software of today. But some of Williams's colleagues set out to model ecosystems nevertheless. Their initial efforts are respected for their ambitiousness. But the results fell short of expectations. Probably the most extensive effort of all around that time was the International Biological Program. As Bruce Alton, an academic biologist in the United States, recalled:

> It started about '65, I think. It was a congressionally mandated program. I suppose it was instigated by a group of ecologists, although

> it would be broader than ecologists, come to think about it, because the International Biological Program was, in fact, an *international* program. Its implementation in the U.S. was given to the National Science Foundation. They did get special earmarked appropriations for the program. The basic structure of it was to take about half a dozen biomes in the North American continent and concentrate efforts in these different regions. There was one point of view when the program began that this was going to grow to a great monster computer model so that you could explain the world. A fellow by the name of George Van Dyne sold that idea to a lot of people, including some congressmen. There was a fair amount of emphasis on computer modeling, particularly early in the program.

Some called this the "whole-ecosystem approach." As Mark Johnson described it:

> I think the whole-ecosystem approach was tremendously successful in some respects and grossly unsuccessful in others. As a kind of awareness creation process, as a teaching activity, it was an excellent way to approach things. The notion, though, that you could take very complicated ecosystems, break them down into their component parts, study the components, and then reassemble them with interacting models that are created through increasingly sophisticated levels of quantitative abstraction—the notion that you could string these together and have them interact and then produce not so much insight, but make accurate predictions about what outcomes will be, that hope really was not fulfilled in the end.

In models, biology saw power and control, an opportunity to organize the frustratingly complex ecosystems that they could theorize but could not empirically explain. This early, failed work nevertheless set the stage for the current explosion in modeling. Yet the tensions evident in these quotes persist today.

Ian Smyth, the modeling optimist quoted above, and others who advocate the whole-ecosystem approach represent one extreme that argues models are indispensable. Without models, scientists will be unable to give policy makers what the latter demand: unambiguous answers to scientific questions. Knowledge. Certainty. No theoretical mumbo-jumbo, but real, solid, scientific direction, what-to-do-next at the touch of a button. While models hold the promise in some researchers' eyes of extended control over the environment, it is plain to see that much of the demand for that control emanates from the institutions that control salmon biology. In 1998 those organizations contributed nearly $2 million for modeling in the Columbia-Snake basin.[10]

Largely, but not entirely, distinct from the optimistic view of models is a perspective that also was alluded to in Alton's subtle skepticism, and more explicitly by Johnson. This view holds that models are best seen as, to use Johnson's terms, "an awareness creation process" and "a teaching activity." Surprisingly, Gary Hite, an expert modeler, agreed. He told me, "You're not using it to predict; you're using the model to help you think, to lead you along and to show you the implications of your actions. All these systems have a lot of interacting pieces, strange time lags. They can do things that you wouldn't think about, counterintuitive things."

George Williams used the same terms: "What computers did was make it possible to explore the properties of systems that were intractable for analytical mathematics. In some respect, I think mathematics has been quite stimulated by the development of computers because it's made mathematicians think of how to generalize what the properties of some systems are. Computers have shown the dimensions, the counterintuitive things that happen. Computers have largely been responsible for stimulating mathematicians to think of chaos theory, for instance, things of that kind, fuzzy logic."

This view argues for models as heuristics. Hite, who helped design some of the most sophisticated models available for predicting salmon runs, also advocates this "middle" approach

> of trying to help you put priorities on things, and then the next thing is to understand what the implications might be: "Given the assumptions about how the world works, and then I do this, what does that mean?" You can start to game. You're not trying to predict things in a quantitative sense, but qualitatively: "If I do this or that, what's the implication?" Again, it helps you understand how a complicated system works. Sometimes quite simple models that are pretty basic and sort of fundamental can tell you some interesting things about, "If I do this now, it turns around and does this to me in the long term." You can say, "That isn't necessarily depen-

> dent on all the structure. Now that I think about it, that's a pretty basic thing." So you can get some real clear messages out of models that don't depend on the model.

Hite's use of the quantitative/qualitative, prediction/understanding distinctions resonates with sociologists, who constantly grapple with these dialectics in debates over the discipline's methodological perspective.

Of course, the irony for modelers is that after quantifying a system, the results still must be interpreted. For those most familiar with modeling, quantification enables understanding, but it does not constitute understanding; quantification is necessary but not sufficient for grasping what is really going on in Nature. The complexity of the systems that are being modeled—which is precisely why they are modeled to begin with—prevents achievement of the quantification ideal of unambiguous exactness. The problem is that biologists are not pursuing the qualitative understanding that connects researchers to the objects of their research, at least not for the most part (although see Chapter Six). The biological approach to understanding continues to have objectivity as its goal, an objectivity based upon the best quantitative understanding that science can bring to bear on a subject.

Whatever movement there might be in biology toward qualitative approaches, I do not want to exaggerate it. Quantification—letting "the data speak for themselves," and the corresponding belief that mathematical models can encompass most of the uncertainty in complex ecological systems—remains the way that all of the biologists with whom I spoke believe salmon biology should be done. They differ with one another only in that some see it as an unrealizable ideal and others as nearly at hand. The move toward qualitative analysis does, however, tell us some important things about the limits of science, both in its methodological abilities and in its capacity to control its own destiny.

Qualitative approaches are prompted in large part by the complexities not only of the ecosystems in which salmon exist, but by the intersections of human social systems with those ecosystems.[11] Pam Mitchell, having noted the unquestioned primacy of quantitative approaches, said, "But I would add to that that this whole concept of the drawdown has thrown us more into that qualitative versus quantitative. The reason that I say that is that all of our quantitative information is based on the conditions that are out there now." Those conditions were subject to changes in public policy, like the drawdown proposal. New political initiatives posed problems that quantification could only partially address. Intuition and empathic understanding, hallmarks of qualitative approaches to interpreting data, must be called upon to fill in the gaps in scientists' knowledge.

When social forces such as public policy in the guise of the Endangered Species Act are brought to bear, they may have the effect of pushing biology beyond itself—of asking biologists questions that they are unable to answer. Biologists are a conservative lot. Lacking data, they are loath to speculate. This means that their models, which explain half or less of the factors affecting adult salmon return, serve only to construct salmon as mysteries. These models have the unintended consequence of highlighting biology's inadequate, partial construction of salmon and its shortcomings as a science when compared to the engineering ideal. Yet the social pressures on salmon biologists to produce a complete salmon are so great that using models as heuristic constructions of salmon is not enough. Certainty and prediction are called for but may not be had, in no small part because the social institutions demanding the certainty are constantly changing the scope conditions.

"ENHANCEMENT": CONTROL BY OTHER MEANS

The foregoing discussion of control of salmon by salmon biologists emphasizes control as something that is done to salmon. The "enhancement" concept includes aspects of manipulating and controlling salmon in this familiar sense, but enhancement also presents a new approach to control not so much of salmon but of the stream, river, and lake systems in which salmon live. Enhancement refers to a range of practices designed to increase the number of salmon found in a given area. They include preserving and rehabilitating streamside habitat to improve the in-stream living conditions for wild salmon—things like placing logs across streams to create pools of slow water. These pools are important for juvenile and adult salmon alike.

But most enhancement activities involve salmon hatcheries. Hatcheries, examined in depth in the next chapter, are used for enhancement purposes such as preserving salmon populations that might otherwise be extirpated (such as when a dam blocks a river), providing salmon for "planting" in lakes, introducing salmon into locations where they are absent (as a result of human or other activity that destroyed the salmon population, or because salmon never existed there), and hatchery salmon are used for "supplementation"—adding salmon from hatcheries to already-existing populations.[12]

Enhancement, then, is a management technique that extends biologists' control beyond the laboratory and beyond the salmon themselves to affect the salmons' surroundings. Politically and economically, enhancement is popular. For example, biologists told me that larger varieties of salmon are planted in streams that historically had smaller-sized fish, thereby providing better sport for anglers. Salmon hatcheries improve the catch for commercial fishing operations in the oceans as well; in recent years the Canadian government embarked on a large-scale enhancement program that emphasized hatcheries for just this reason.

However, salmon hatcheries and the increased control that is inherent in them create further problems for salmon and for society. This is true for nearly all large and complex systems, whether they are fish hatcheries, chemical plants, or nuclear power stations.[13] A provocative instance of such unintended consequences comes from Canada, where the expected returns of hatchery-bred adult salmon did not transpire, causing considerable consternation. As Ian Smyth explained:

> There are concerns about enhancement, because if the fish aren't surviving, how come there are so many of them? The bottom line is, if you start looking at how many enhancement activities there are, even with low survival in all of these things you still end up with a lot of fish. In the Straits of Georgia, I honestly can't name any significant river system without an enhancement program in it. The proliferation of this stuff is incredible. They're all different scales, but it means that people are in there doing things all the time, that there aren't any truly undisturbed or natural systems there.
>
> *Q. Does that trouble you?*
>
> SMYTH: Yeah, that does trouble me, actually. I think that the impacts will depend on what they're doing there, of course. But the fact that we don't have any systems that we're just letting work itself through the changing environments and this sort of thing, that does bother me. People that I talk to say, "Well, what if they aren't going

> to survive without it?" I say, "Actually, I don't see any reason to suspect that they couldn't survive, unless we are not responsible and we destroy all the habitats."

Implicit in Smyth's comments is that this wholesale enhancement activity is an experiment and that the hallmark of the experimental method—the existence of a "control" that does not receive the experimental treatment (in this case a control would be a stream without a hatchery on it)—is completely absent. No one knows or can know the extent of enhancement's effects on the streams and rivers in question because science been forced into the backseat by policy and economic concerns.

Smyth said that his primary anxiety was that wild salmon were being affected by the proliferation of hatchery fish. One of the most extreme examples of this occurs in "mixed-stock" fisheries, ocean areas where hatchery and wild fish swim together and where they may be simultaneously caught. Mixed-stock fisheries hold the potential for wholesale destruction of fragile wild salmon runs, according to biologists. They only add to the difficulties faced by wild salmon from development, dams, logging, cattle grazing, pollution, and other human activities.

Enhancement programs may have further unintended ramifications for society. In recent years the looming threat of a "salmon war" has hung over the commercial fishing seasons in the near-coastal waters of the United States and Canada, in part a result of British Columbia's hatchery enhancement efforts.[14] The most monetarily valuable salmon, the sockeye, spawn up the Fraser River and its tributaries in huge numbers. Most of these fish are wild, but enhancement has been successful, and an increasing proportion of these "new" fish are homing to Canadian hatcheries. Some of those salmon, however, swim through U.S. waters on their way to their spawning grounds (or, in the case of hatchery fish, on their way to being spawned, since this process is done by hand by humans), and the U.S. fleet has claimed the right to catch them. The Canadians argue that most of those fish should be caught by Canadians because the Canadian government paid for the rearing of the fish. In turn, the U.S. government says Canadian fleets catch too many salmon bound for U.S. waters, including fish from decimated wild populations.

In the summer of 1994 the Canadians retaliated for what they felt was intransigence on the part of the U.S. government in the dispute by charging Alaska-bound U.S. fishing boats a C$1,500 passage fee when they sailed through Canadian waters. Had the hatcheries not been built, the dispute, in which millions of dollars of salmon are at stake each year, likely never would have arisen. But there would not have been as many fish to catch, either. (A detailed discussion of the Salmon War may be found in Chapter Seven.)

Assessing Enhancement

Enhancement implies that Nature is incomplete and that humans can measure—quantify—the difference between what is and what ought to be in Nature. The initial difficulty in this calculation comes in determining how the ideal state—complete Nature—is to be measured. This is essentially an effort at controlling uncertainty, and a remarkable example comes from a conference presentation by biologists who worked for a Native American tribal hatchery. Their intention was to "fully utilize" the carrying capacity of a portion of one river system where coho salmon had spawned for millennia (carrying capacity is "the maximum population of a given species which a particular habitat can support indefinitely").[15] The scientists captured wild fish, bred them in the hatchery, and released what they computed to be the proper number of fry into streams to "maximize yields" so that the stream operated at full carrying capacity. The scientists reported their results, saying,

> There was not as large an increase in smolt production as had been anticipated. . . . In cases where natural fry seeding was low, increases have resulted. In cases where natural fry seeding is average or higher, no additional smolts are produced, apparently, especially in tributary habitats. When these conditions occur in the tributaries, it suggests that smolt carrying capacity is achieved on the average. However, carrying capacity does appear to vary considerably in these streams. In situations where seeding levels are adequate to achieve or nearly adequate to achieve carrying capacity, [they] have resulted in a loss of naturally produced fish through replacement. . . . On a systemwide basis, fry supplementation . . . would likely result in benefits to the river and to the fishery less than half the time.

The scientists' efforts achieved few of their intended goals, and in some instances the practice harmed wild salmon. Yet again, the intended physical construction resulted in unintended consequences that actually harmed the revered, and often dwindling, "natural" (wild) fish populations.

As this example shows, enhancement, including supplementation, often depends for its success on the quantification of variables that are difficult, perhaps impossible, to quantify, such as "stream carrying capacity." Also difficult to quantify, and even to identify, are the relevant human behaviors that may affect carrying capacity. For example, biologists told me that they were concerned that global climate change (the "greenhouse effect") may be reducing ocean carrying capacity for salmon by disrupting the food chain. Within salmon biology proper (interactions between biologists), salmon biologists are effective salmon biology-sociologists; outside of the discipline, however, biologists are no better at predicting the impacts of human behavior than are sociologists, the hopes of those in both disciplines notwithstanding.

THE INTERCHANGEABILITY OF SALMON

The final type of control that salmon biologists exert over salmon is what I term *interchangeability.* Interchangeability is a multifaceted concept. It has three different senses, two of which are relevant here. The first is *interchangeability of salmon.* This refers to the perceptual and behavioral treatment of salmon as "all alike"—as biologically indistinguishable from one another. Second, the *interchangeability of species* refers to the treatment of entirely different types of animals—say, salmon and bats—as the same for some purposes. The third form, *interchangeability of biologists' abilities,* is examined in Chapter Six.

Most biologists characterize the interchangeability of salmon as a disturbing problem. Yet it is a problem of the biologists' own making. Some assert that as a practice it is virtually nonexistent today, but others are less sanguine. Jim Winston neatly summarized this phenomenon when he angrily recounted that biologists in the past "didn't understand homing, stock specificity. We're still fighting that in the scientific community today. There are scientists that don't put a great deal of stock in a population's relationship to its unique environment. That's demonstrated by the fact that we transplant populations from river to river without any regard for the uniqueness that they must require in order to sustain themselves. Science, at least in how we view science, has not been that well practiced in fisheries." This kind of interchangeability was alluded to above when I noted that some stocks or strains of salmon—genetically unique groups—have been introduced at anglers' request as part of enhancement programs. Planting larger-sized fish of a species into areas where smaller-sized fish of the same species have lived historically is one of the most common forms of salmon interchangeability.

Species interchangeability is the notion among biologists that, for scientific purposes, abstract phenomena are what are of importance, not the species in which the phenomenon occurs. This idea was expressed by a Canadian who has published research findings on other species. He observed:

> The basic principles are the same. I think you have to get in the state of mind of a researcher who is thinking about a phenomena. If you're thinking about the population phenomena, why do populations fluctuate in abundance, what regulates their abundance, it doesn't really matter whether you're working on insects or moles or fish—the basic principles are the same. You can take a course in ecology, and if you've got a good grasp of the principles, you can apply them to any animal. It didn't bother me that I worked on moles rather than on whitefish or cod. As I say, underlying it all is the same basic message.
>
> That gets off into a whole area about what sort of education you should get at a university. I remember one of my friends on the East Coast remarked once, he was so infuriated. He had hired a kid who had gone to some university in the southern States somewhere to do research on blue crabs. Then it turned out that the money for the blue crab study disappeared, so he told him he was going to work on oysters instead. The guy said he couldn't do research on oysters. My friend said, "Why not?" and he said, "Well, because I never took a course on oysters." I think I could get up to speed to work on population dynamics of elephants in six weeks. I've got to know the birth rate, the death rate, predators, parasites. It's the same bag of tricks. It's just a matter of learning a different vocabulary and having a little bit of background, natural history and so on.

The same biologist went on to say that "if you're interested in a phenomenon in population biology, whether it's competition or predation or whatever, it doesn't really matter what analogue you use—moose, elk, fish. . . ." The particulars of one species or another are of little interest. By uncovering their similarities one can move ever closer to positivism's goal of universal truth. Species interchangeability is an effort to fulfill that socially sanctioned scientific ideal.

These two types of interchangeability exhibit both explicit and more subtle forms of control and power. For example, it is an act of substantial control to transfer members of one population of salmon perhaps hundreds of miles from their historical habitat into new areas, and doubly so when that act of planting displaces other fish, as often has been the case. Similarly, for a discipline such as biology to have developed concepts like those of population dynamics that are transferrable from one species to another without alteration demonstrates the power of its theoretical constructs. Biologists impose a workable mental ordering on much of the living world.

More subtle, yet crucial in facilitating the interchangeabilities, are the norms that justify, for example, the reconstructing of a kokanee salmon (a landlocked sockeye salmon) from Babine Lake in Canada as a Wallowa Lake, Oregon, kokanee. Babine Lake is lower in elevation, farther north, and nearer the ocean than is Wallowa Lake, yet transfers of kokanee from lake to lake have often occurred. Similar professional expectations demand of biologists undertaking "good biology" that they be able to move seamlessly from studying blue crabs to studying oysters. Identification of animals with a place, and identification by biologists with animals, is absent in the classical fisheries biology model. The norm is that of objective science, which conceptualizes the workings of organisms as machinelike and directs that those workings are all that matters. If geography or some other factor can be justified as theoretically important, then it, too, should be added to the equation. Otherwise, it is irrelevant.

Many of my research participants pinned clinical-looking salmon identification posters, replete with genus and species names, onto their walls (and I tacked up a set of my own above my desk to help me understand a bit more about their world). Some participants wore belt buckles depicting the fish. Others prominently displayed photographs of their favorite research sites. Yet I was surprised at their lack of emotional identification with the fish. I asked nearly all of them why they studied salmon. A few biologists grew up fishing for salmon or living in the Pacific Northwest, and their childhood exposure to the notion that salmon were important to the region had a lasting impact on them. Many did not come from the area, however, and some ended up studying salmon almost by happenstance—the money was there or it was a good job, I was often told. A "good job" meant that research facilities were adequate and the "species was interesting."

This all leaves one with the strong sense that biologists construct salmon merely as one species among many. Like all biological organisms, salmon harbor numerous mysteries, but for most biologists they are mysteries of the mind, not of the heart. After all, emotions are dangerous things. They cloud judgment and lead one to take sides, something the template says scientists ideally ought never to do. Some of my research participants told me of the fishing trips for salmon that they took as children and the sense of connection and even awe that they felt toward the fish. Those experiences set them on trajectories that have shaped their lives. But when they became immersed in the intense world of graduate study, they were resocialized to biology's more hard-nosed, positivistic ends: exploring theories, asking the right questions, collecting quality data, doing good science. Salmon may be beautiful, fun, and delicious for those who both fish for and study them. As professionals, however, many biologists see salmon as only one of any number of organisms that they might study.

Biologists construct salmon species as interchangeable with one another. In many ways they are no different from organisms that appear wholly unlike fish to the nonbiologist's eye. Salmon become part of an undifferentiated, homogenized Nature. From salmon to elephant, the parts are the same, though the wholes may look and feel as distinct as night and day and may live in entirely different worlds. The result is an oddly commonplace salmon—a mechanical, schematic, engineered fish—to many of those who know the most about them.

SALMON BIOLOGY AND CONTROL OVER MANAGERS

Nearly everything that salmon biologists do is controlled by social factors, including the control that they exert over salmon and Nature. But biologists also control or seek to control other social groups in the course of their work, and this plays a central role in their constructions of salmon. Sociologists would expect that all salmon biologists, being salmon biology-sociologists themselves, would take pains to attend to the human aspects of their work. But this is a more immediate concern for salmon biologists in applied areas, if for no other reason than that their work is intended for direct use by social groups, such as hatcheries or fish-processing companies, and so they directly confront the social ramifications of their endeavors. Nevertheless, those who undertake basic or long-term studies also demonstrate an awareness of the importance of control over humans for the success of their efforts.

BIOLOGISTS AND MANAGERS

The bulk of Ian Smyth's research is applied, and this has brought him into contact and conflict with a job type that many biologists noted with contempt: management. Smyth explained some of the dynamics of the contentious biologist-manager relationship in Canada, noting the uncomfortable gray area that has been created between fisheries managers and fisheries scientists.

> The way we try to differentiate our work between the two is that the fisheries scientists look at the ecosystem limits to production, understand the population dynamics of the species. Managers, on the other hand, are managing the people in the fisheries. They're more of the social managers. Because the managers have this real time: Every year they have to go through this management situation: how many fish are coming back, how many can be allocated to what user? They need a lot of information and they need it very quickly. To try and address that, they brought in biologists that act like fisheries scientists. So they're doing a lot of science-type activities. But the fisheries sciences group doesn't have any control over what they do. The managers also get quite a bit of money, of course.

These "social managers" set regulations controlling when fishing can take place, the number of fish that can be caught, the fishing gear that can be used, and related issues.

Compared to applied salmon biologists, managers are even more focused on economics; frequently, this is expressed as "maximizing the yield" of salmon that are "harvested" in a particular fishery (a location where fishing takes place, be it on the high seas, a river, or a lake). This is particularly true in the management of commercial salmon operations. In commercial fisheries, I was told by a Canadian biologist, "you'll find a lot of information in seasons, very detailed procedures to estimate your catch as you're going through your season. You always try to catch the last fish because it's the thing to do, the optimum thing. You're always very close to the edge. And it doesn't work to a large extent. I think native people whom I've talked to are more willing to take a more flexible view each year than pushing it so close to the edge. So they may be open to different management styles." An American said something similar: "We take away, we exploit so greatly this resource and do not reinvest back into its protection from an ecological point of view and from a population dynamics point of view. We manage this resource on the 'ragged edge,' a term we often use."

Managers are the ones who establish that ragged edge. The "escapement levels" they set are especially important. These are the numbers of salmon that must go uncaught to ensure a sustainable population of fish from generation to generation (those are *fish* generations, of course, about three to five years). This means that managers exert substantial control over those who fish and how much they can catch. As a result, the managers are politically powerful, and as such they occupy highly political roles. They may be exposed to intense pressure to adjust escapement levels to satisfy the wishes of commercial fishing fleets and politicians. Indeed, several biologists told me that ragged-edge escapements are based more on politics and economics than they are on "sound data."

Political pressure on managers can come from many fronts. In 1998, the Canadian Automobile Workers Union, which represents twenty-six thousand fishery employees, wrote, "Close to a million surplus late-run sockeye that could have put another $15 million into commercial fishermen's jeans and several more million through processing plants will likely not contribute anything to the beleaguered fishing industry. Fish managers didn't even know they were there until they'd gone past the Mission hydroacoustic facility and were on the spawning grounds in the Birkenhead River and Weaver Creek in the lower Fraser watershed."[16] It is doubtful that managers will be allowed to make the same "mistake" again in the future.

A contrasting example of just how nonscientific escapement decisions can be occurred in 1994 with the Adams River sockeye salmon run. The Adams, a tributary of the Fraser in southern British Columbia, has the most concentrated run of salmon in the world, with 4 million or more fish returning to a section of river only nine miles long. Every four years the run peaks, and when I witnessed the 1994 run at Roderick-Haig Brown Park near Salmon Arm, it was spectacular. Sockeye, bright red and green in their spawning colors, fought one another for nesting sites or for the right to mate, splashing and thrashing all the day long. Those that had already spawned lay dead and dying amongst bright fall leaves.

The scene was simultaneously thrilling and depressing, so much life and so much death in such a small place. Yet newspaper accounts reported that the run was disappointingly small, only a couple million or so fish in a peak year. An investigation ensued, and although the cause remains something of a mystery, poor management of the commercial fishing fleet was clearly one factor in the relatively small run. Indeed, the whole thing was nearly a disaster. As the *Seattle Times* reported four years later, "In 1994, biologists believe that competing Washington and B.C. fishermen nearly fished the Adams River sockeye to the brink of extinction. A subsequent study concluded that fisheries officials closed down fishing in the nick of time; if fishing had stayed open another 12 hours, the run would have been depleted."[17] On one side of the all too ragged edge is escapement and management success. On the other, managed extinction.

The biologist-manager relationship is actually one of mutual social control, for managers wield considerable control over salmon biologists. However, this control over biologists differs from that exerted by structural interests in that it often is *control by social closure.* Smyth's perspective, quoted above, is that rather than admit certified fisheries experts—who, bureaucratically or in terms of social status, are the managers' peers—into the Canadian regulatory process, fisheries managers use their economic power to hire nonfisheries biologists "that act like fisheries scientists." This ensures that the managers will have control over scientific studies, and therefore control over information, because the nonfishery biologists are answerable only to the managers, not to a profession with its own standards. So managers close off their operations to fisheries biologists in order to get the results they want, according to salmon biologists.

Extending the prior discussions on research and the power of funding agencies, we can see that controlling research may be crucial to management's relationship with its primary constituency, be it commercial fishing fleets or recreational anglers. This creates conflicts with biologists, who seek control over managers because they argue that they "know the resource" better than nonscientifically oriented, politically motivated managers, and thus their research is superior to that of any other group.

Jim Winston, whose research interests are distinct from Ian Smyth's and who works in the United States, also viewed managing agencies with contempt. He told me: "We know so little about the most important species that we deal with . . . , and that's evident in how we manage them. It's not that they're managing in ignorance, that they could care less; it's just that they are not managing with the best understanding of the animal that they could have. So they cannot respond to the needs of the animal. I think that probably because I see a lack of it; life history in general, and specifically migratory life history, is extremely poor."

In the United States, as in Canada, strong political constituencies affect managers' "understandings." Managers appear to construct salmon narrowly, as a commodity to be maximized or as a political problem to be minimized. As we have seen, biologists' dominant constructions are similar, but biologists argue that their outlook is qualitatively different. Biologists claim to bring an understanding of the animal to the fore. For this reason one would expect that their "ragged edge" might not be quite as ambitious as managers' if the decisions were left to them.

An example of managers' minimization of potential biological problems, and the contrasting concern shown by biologists, comes from Pam Mitchell. She was more blunt than most when asked about managers. In discussing the Snake River reservoir drawdown proposal and the research needed to determine its potential impacts, she pointedly blamed ignorance for what she saw as poor management decisions:

> I would say the politicians don't want additional research because they're afraid that it might cause somebody to think twice about supporting the drawdown position. But amongst the biologists who support the drawdown, I honestly think it's ignorance. They think they know all they need to know. . . . When I get some of those people aside one-on-one and say, "Look at this and what about this?" they say, "Oh, I didn't realize all that." Then they go back to

their agency folks who once again say, "We know that." I've seen that happen again and again. I just think, "You guys aren't helping the fish, here!" Of course, some of it is so difficult to research. But ignoring the problem doesn't make it go away.

In this case, biologists who were not Mitchell's peers but who were in management organizations and in a position to direct her research were threatening to bring premature closure to the problem of the impact of drawdowns on the salmon.

Owing to managers' economically and politically powerful positions, biologists have effectively been kept out of decision-making processes where, they assert, they and their studies belong. Many salmon biologists perceive themselves as having an interest in managing—controlling—human behavior toward salmon. It also appears that they find it difficult to wrest that control from others, and they even have a hard time getting their studies to be considered as integral to management decisions (read: control over humans).

This observation runs in the face of assertions, like those by Bruno Latour and others, that science is the source of new power in society.[18] More realistically, given events like the creation of an apparently redundant science branch in Canadian fish management agencies, *certain* science—that which reinforces and upholds preexisting dominant social interests—is a source of *lasting* power and control in society. As one biologist observed, "Policy makers evaluate the science, but the science rarely carries the day." In this milieu, biologists' constructions emerge in an oppositional setting. They come into frequent conflict with managers. The biologists have data, or have identified data that they say are missing and need to be gathered. The managers—under the gun to set escapement levels and to make other decisions that will have political, economic, and other social ramifications—often appear satisfied with whatever information they have.

The result is that biologists construct themselves and the salmon in opposition to the managers. Biologists' self-construction is as the benign overseers of the salmon. They stand between controlling managers and the helpless, at-risk, threatened, and endangered salmon. Salmon science becomes the salmon's white knight, its savior in a rapacious world. In time we will see that some salmon biologists are taking this construction of themselves to an extreme by attempting to use the power of their profession against the very interests that now support them.

CONCLUSION: SALMON AND BIOLOGY TRANSFORMED

With the addition of the concepts discussed in this chapter to those in Chapter Two, we have a strong sense that control of salmon is central to what salmon biologists do, and in its multiple incarnations control profoundly affects biologists' construction of the fish. Control is omnipresent and multifarious. Biologists exert control and they are themselves the objects of controlling, powerful, rationalizing influences in society and within their own profession.

Salmon research depends upon the extent of control that biologists can exert over salmon. However, this is as much a social process as it is a biological one. Conceiving of salmon as existing in a system and then bringing conceptual closure to that system is crucial to this process, and this rush to bring engineering-like certainty to salmon biology is spurred by a shift in the macrolevel forces shaping biologists' work. Science is not the primary determinant of research needs, and it is unlikely that it ever has been. But today the political/governmental and economic forces that have played such a large role in North American salmon biology from its inception impress upon scientists a new sense of urgency. These pressures lead to a narrowing construction of salmon. *Even to biologists the salmon become embodiments of public policy and tools for economic gain.* Most revealing sociologically is not so much the idea that salmon have come to be mere surrogates for political or economic power in the biologists' eyes as that salmon have become political and economic objects to powerful social forces, and in turn these have transformed biology.

The transformation occurs by controlling resources, especially by funding highly specific projects directed at an "actual pressing question of the day." The effect is to control the research issues of interest to biologists, and this affects their constructions of the salmon. Governmental and economic entities socialize biologists to view salmon in utilitarian terms. Salmon long have been constructed as important economic entities, whether the society was pre-Columbian or industrial.[19] But never before have those who study salmon been compelled to adopt a policy and economics-based construction in order to pursue their work.

What results is a highly rationalized salmon, and the engineering ideal type serves as the touchstone for salmon biology in this regard. After all, what are "systems"? Are they not supremely rational: that is, calculable, ordered, precise, efficient, certain, and above all *knowable?* Systematizing salmon yields a known biological entity, one that performs on cue with little or no margin of error. Such exactness, were it ever to be achieved, would serve human ends quite nicely. Salmon could be caught anywhere; even today their oceanic wanderings are largely a mystery. There would be no more "ragged edge" escapements; rather, salmon could be fished to the last, leaving just enough escapees to ensure a precise harvest years down the line. And salmon biologists would attain the status of engineers; positivism's dream of a perfected salmon—a salmon that obeys laws discovered (read: created) by scientists—would be at hand, and with it would come the power, prestige, and privilege enjoyed by engineers.

One problem, though: people. Whether they are an anxious Northwest Power Planning Council pursuing "adaptive management" and thus shifting the biological playing field for salmon every few years by authorizing drawdowns and other experiments, or whether they are agency managers wanting to field their own team of scientists, social groups make a positivistic, rationalized construction of salmon impossible to realize.

And a second problem: salmon. The following chapter, which explores the "tooling" of salmon in hatcheries, supports the *Jurassic Park* contention that "nature finds a way" to thwart society's best-laid schemes. Salmon do this precisely because society cannot extend its control to everything. No system is truly closed and impervious to outside influence, despite what engineers and some salmon biologists like to believe. And despite Max Weber's worst nightmares, no society can be completely rationalized.

4 Thinking and Making Salmon

TIM ARMSTRONG STRUCK me as the oddest of salmon hatchery biologists. As a graduate student he had spoken out against the construction of a dam that was, he felt, certain to bring destruction to one of the Pacific Northwest's premier salmon and trout streams. Years later, he found himself in charge of the biological side of a salmon hatchery that was constructed as mitigation for the habitat lost when that same dam was built. Hatcheries rear fish from generation to generation. The idea is that—after fish have been spawned and reared in hatcheries, then released to roam the oceans—some of the adults will escape harvest and return to allow biologists to spawn a new generation. A "mitigation hatchery" like Armstrong's serves as recompense for the loss of spawning, rearing, and fishing grounds due to dams.

Armstrong, a large man wearing a green uniform and speaking in a soft drawl, never wanted to see the dam or the attendant hatchery where he now works constructed. Yet once they were in place, he was eager to be a part of making the complex work for the salmon, driven as he was by a sense that only someone who cared about the fish as much as he does could do the job right. Perhaps because of his long opposition to the construction of hatcheries—and in particular to what is now *his* hatchery—he has remained outspoken about their shortcomings. He differs with those who ardently defend hatcheries because he readily, unflinchingly acknowledges hatcheries' shortcomings as identified by the dominant biological model. In particular, he says that it is the technology of the hatchery that creates these problems. Around a large conference table at the hatchery he told me,

> You have to realize that we have changed the fish in a hatchery. We've changed the success of fish in a hatchery. . . . We grow a smolt that used to take two to three years to smolt, we do it in a year. So

> no matter what we do—we try to take the eggs over a period, we try to match them—but no matter what we do, those first eggs that we take always have somewhat of an advantage. If we start taking eggs in, say, February and we take them through the first of May, those that we take in the first of February are going to have more advantage on growth. . . . The bigger, the higher success rate. So you perpetuate stuff like that over so many generations. And we're not doing it intentionally. We're trying as best we can at making some management decisions to try to keep later stocks and keep them going. But it's just the nature of what we're doing here, the technology, that those fish have a better chance, those earlier-spawning fish. So over a few years, and this happened here, we got the peak of the run that's now in the first week of April in this hatchery. Well out here in the wild the first week of April is probably not the peak. We're probably moving them a month ahead.

Both advocates and opponents of hatcheries said similar things. Science and technology change salmon; the fish figuratively and even literally become technologized. As Arnold Arluke and Clinton R. Sanders wrote in a general discussion of the social construction of animals, "To become tools . . . their animal nature must be reconstructed as scientific data or food."[1] Ultimately, salmon themselves become scientific and technological processes.

As previously noted, Max Weber observed that the edifice of rational social life is constructed upon technological and scientific progress.[2] Together, economics, politics and government, bureaucracies, technologies, and science threaten to rationalize all of social life, and all that society controls. This includes salmon, which have been the objects of these rationalizing forces for more than a century. They render salmon as products, commodities of a modern industrial social system, and in the process these rationalizing impulses imbue salmon with meaning.

Nowhere is the technological rationalization of salmon more obvious than in fish hatcheries. Less than a century ago hatcheries were seen as providing the possibility of endless food for all Americans. "As a consequence," wrote one biologist-historian, "the federal government embarked upon a vigorous, and apparently completely futile, program of fish culture which persisted for more than sixty years."[3] Utter failures in the past have not deterred hatchery advocates, however. Biologists say that those early disasters were largely the result of incomplete mimicking of Nature at hatcheries, yet even today some of those old practices are pursued.

The process of growing salmon in hatcheries is complex. Essentially, though, it begins when eggs are cut from "ripe" female salmons' bellies and milt—sperm—is squeezed from selected males. The two are mixed in a bucket by technicians wearing rubber gloves, then the fertilized eggs are poured into plastic incubation trays stacked one atop another. Water, drawn from nearby streams and heated just enough to speed up the growth of the eggs yet not kill them, constantly runs through the trays. When the "parrs"—fingerlings—have emerged from their eggs and have absorbed their yolk sacs, they are transferred to huge tanks holding hundreds of gallons of water and thousands of pinkie-sized salmon. Until they are released from the hatchery months later, the fish are fed by hand or by an automatic feeder. Technicians eventually move the young salmon from the tanks to "raceways," concrete ovals with a concrete island in the middle resembling miniature horse-racing tracks for fish.

Some of the young salmon may be experimented upon, and they may have a fin clipped off for identification purposes. Other means of identification are through exposure to chemicals that imprint special codes into their bones and even branding, using a liquid nitrogen branding iron. Eventually they will be released from this artificial, thoroughly human-created environment into what surely must be a foreign world for them of rivers and streams. There they may carry with them microchips or coded bits of wire to allow biologists to gather experimental data.

Throughout their lives these hatchery salmon are tooled—worked, manipulated, altered—and through this tooling they become society's instruments. As Tim Armstrong's quote at the beginning of this chapter indicates, hatchery salmon are not wild or Natural fish in many biologists' eyes. Society's interactions with hatchery salmon are quite different than those with wild salmon. Hatchery salmon are human-created in a literal, physical sense, and new developments in genetic engineering may ultimately allow salmon biologists godlike powers of creation, changing hatchery fish in elemental ways. To nearly all who have interactions with hatchery salmon, they are *less than Natural* because of society's technologization of them.

But wild salmon are tooled as well. For example, they, too, may be used as research subjects, and both they and their hatchery cousins are exposed to human artifacts from dams to pollution to the barges that are used to ship many of them downstream and past danger (after being captured inside of dams and flushed down long pipes to holding facilities). Why do distinctions between classes of salmon persist? How does tooling salmon alter our constructions of them? How has placing salmon in hatcheries and changing their genetic structure given meaning to the fish, and what are the meanings that result?

COGNITIVE AND PHYSICAL CONSTRUCTIONS

To begin with a basic observation: *salmon are manipulated to serve human ends.* If this were said about livestock or perhaps even goldfish, the statement would appear trite. After all, cows, lamb, and aquarium denizens give us food, 4-H honors, company, and visual distractions. And if the statement were generalized to read "nature is manipulated to serve human ends," a casual reader would be justified in labeling this assertion "obvious." Of course we use Nature. Every human society does. However, my sense is that this observation needs to be made explicit because it is one of the basic taken-for-granteds of life in modern societies that, to some extent or other, Natural entities are used, manipulated, and controlled for human ends. This assumption is so commonplace as to be virtually invisible, unnoticed by most of us. When spoken, it may be obvious, but most of the time it is hidden. So few of us are actually involved in working the land and oceans or working with the products from seas, oceans, mines, and forests that it is easy to take them for granted.

Only at momentous times do we consider the extent of our interactions with nonhuman entities. Examples of this include the Arab oil embargoes of the 1970s and the Gulf War in 1990–91, when oil, a vital Natural resource in the West, suddenly was in short supply (although it was made so because of social forces). Similar examples of the inconvenient times when Nature reminds us of its presence include "natural disasters" such as hurricanes, earthquakes, and volcanic eruptions. When these events occur, whether through some sort of social action or inaction or through forces outside of social control, human interactions with the nonhuman world suddenly are made problematic. And it is at these times that they attract attention from scholars, politicians, and the public at large.

But it is the commonplaces, not the rarities, that interest us here. This chapter seeks to explore the simple, commonplace proposition that salmon are manipulated to serve human ends. Why bother to do so? One reason is that sociologists have a long history of exploring society's taken-for-granteds in an effort to discern "what is going on here."[4] John Van Maanen has written that qualitative sociology examines society's ontological status and that doing so illuminates the mundane realities of our worlds.[5] What is there to be discovered about society by exploring human manipulations of salmon?

The social construction of salmon actually proceeds at two levels: cognitively and physically. *Cognitive constructions* emerge from the mental perspective that humans have power over salmon and may manipulate them; they are ensconced in texts and discourse. *Physical and behavioral constructions* point to society's control over salmon and are embodied and enacted—as opposed to spoken or written—sources of meaning as found, for example, in a society's technologies (such as dams, fish hatcheries, and fishing gear), in its acted-out rituals and ceremonies (a Native American tribe's first salmon feast, a public hearing on a proposed new dam and its effect upon fish), and in day-to-day nonverbal interactions (including human-to-human and human-to-salmon behaviors such as dissecting fish in a laboratory or catching fish by anglers).

Dividing the social construction of Nature concept into its cognitive and physical constituents helps in identifying the varied social forces that construct salmon, and for the purposes of this chapter the rationalizing forces in particular. Following a brief overview of each of these approaches, I hope to demonstrate the impact of rationalizing influences in interactions between humans and salmon, emphasizing an agricultural model, salmon hatcheries, and dam bypass systems for salmon.

COGNITIVE CONSTRUCTIONS

Two important sociological observations emerge from cognitive construction's emphasis on language and thought. The first is that this concept is malleable; it allows for social change. Linguistic and textual meanings are generally stable, but they do evolve in modern societies, and theoretical concepts should address the inevitability of change. Second, changes in cognitive constructions most often occur in response to macrosocial structural change. The agents of social change may take many forms: technologies, social movements, and more diffuse collective behaviors are just some of the examples relevant to this study.

Perhaps the most obvious example of this concept is the cognitive construction of salmon as food by many cultures. Salmon as a source of sustenance was probably one of the earliest meanings attributed to the fish, and this is likely what gave rise to salmon as important religious symbols.[6] The traditional Native American spiritual connections with salmon were supplanted as the dominant construction by society's industrial meaning of the salmon: as entities whose sole value was to serve human ends.[7] Today the fish are seen from a utilitarian and mechanistic perspective, both by salmon biologists and by society more generally, and this view is inculcated through the centers of fisheries education. As Mark Johnson, a Canadian biologist, told me in speaking of sockeye salmon and why he researches them, "It's a very satisfying species to work on from a whole range of perspectives. One is it integrates a lot of the background experience that I've built up to date and makes best use of it. Secondly, from a commercial and practical standpoint, they're the most valuable of the Pacific salmon species. They generate the most economic rent back to the general population." Like indentured servants, the salmon must pay their own way in our society, their worth being based exclusively on their dollar value.

Some salmon biologists speak in unambiguously mechanistic terms, as when Tim Armstrong said, "Now in chinook, they seem to smolt as they're going down river. There's something about moving in water and starting the process that seems to kick off some mechanism, at least that's the way it seems." Others were more subtle. Neil Self commented, "Yes, they're beautifully fueled. The analogy some people use is they're like rockets. Take one that is aimed to go two thousand miles: fuel to mass balance is very different than one that's aimed to go six thousand. And the way they move energy around is delightful." And even at the cutting edge of salmon biology, among those who stress the role of genetics in their work, similar constructions can be found. Another Canadian told me, "But certain conditions will suddenly trigger the genes—it might be application of a herbicide or a pesticide, or whatever, and maybe 90 percent of the population may die. But those 10 percent, they'll switch and survive." Genes in salmon are seen as biological light switches. This mechanistic metaphor has been in the vernacular since early in the Industrial Revolution. In the case of salmon, cognitively they are systematized; they are not organic wholes but are instead reduced to so many separable parts. They are genes being triggered and calories being channeled.

Arluke and Sanders wrote that, in order for animals to be tooled, they "must be deanthropomorphized, becoming lesser beings or objects that think few thoughts, feel only the most primitive emotions, and experience little pain."[8] This reductionistic cognitive model was expressed in the philosophy of René Descartes, who argued that animals were "irrational machines."[9] Alternative cognitive constructions of salmon, such as those embraced by Native American peoples before industrialism overwhelmed them, and still held by some tribes today, are well documented, and some of the participants spoke of these. But the dominant view is a far different construction of the salmon than those of Native Americans, as it must be in order for science to pursue its ends, specifically to identify the causal mechanisms in all that salmon do.

Marjorie Grene wrote that the ultimate reductionistic "view that life is 'nothing but statistical variations in the gene pools of populations' " is at hand.[10] Because of science's longing to discern cause and effect in salmon, especially "ultimate causes," some salmon biologists argue that it has lost all sight of understanding salmon. However, as Carolyn Merchant observes, an ecological way of seeing begins to recover some of that sense of empathy, of *Verstehen,* that is lost in the Cartesian cognitive construction.[11]

PHYSICAL CONSTRUCTIONS

In order to completely rationalize the salmon in industrial cultures, they must be manipulated. The salmon must be tooled, be worked so that they become a cog in industrial mechanisms. Jack P. Gibbs, who wrote so extensively about control, defined all technologies as "inanimate."[12] In contrast, I will argue that society's treatment of salmon demonstrates that, in certain circumstances, living organisms can be properly viewed as technologies. Specifically, salmon have been turned into research instruments. This is not unique—members of numerous other species have been fitted with radio collars, for example. Such acts reconstruct any animal not just as an object of scientific research but as an instrument for research.

Virtually all of the interactions our society has with salmon are technologically mediated. Phenomenologist Don Ihde refers to such interactions as "embodiment relations." He writes that these "simultaneously magnify or amplify and reduce or place aside what is experienced through them."[13] For example, whenever any social group imbues a salmon with meaning—whether by carving a likeness of it into a piece of wood, producing it in a fish hatchery, or inserting a microchip into its back to study it—it is no longer the being as simply experienced. By magnifying the produced salmon through our hatcheries and other technologies, we reduce the experience of the salmon and reduce the salmon as a self-important entity.

Thus, we should not merely understand the tooling of salmon as altering their meaning, but as an imposition of new meaning upon the fish by creating whole new salmon. This is an important statement about both meaning and society's power to redirect meanings. Technological creations of salmon, and the cognitive constructions that allow us to alter the salmon experience in the ways that we have, are the subject of the remainder of this chapter, where I consider the construction of salmon in the Pacific Northwest via technologies such as salmon hatcheries and the use of high technology means to monitor and to re-create salmon.

SALMON HATCHERIES AS POLITICAL-ECONOMIC INSTRUMENTS

SALMON HATCHERY TECHNOLOGY

Salmon hatcheries play a huge role in the overall "production" of salmon in the Pacific Northwest (I discuss the meaning-laden *production* concept in the following section). There are 128 federal, state, and tribal hatcheries in Washington State alone; these send 350 million young fish annually to the ocean. (Ironically—perhaps incredibly—a nineteenth-century biologist asserted that he "had found only two suitable hatchery sites in the Columbia River system.")[14] One of my research locales was "Northwest Salmon Hatchery."[15] Among the largest fish hatcheries in the nation, Northwest Hatchery produces more than 2 million salmon each year. Its scale is what separates Northwest from most other hatcheries; aside from that, they all have similar characteristics. There are buildings (or sections of a single building) for administrative personnel, for scientific studies, for "mating" the adult salmon that return to the hatchery, and for rearing young salmon.

But the visual and practical centerpiece of any hatchery is the "rearing ponds" or "raceways." Hatchery salmon do most of their freshwater maturing in these odd-looking pools. The raceways vary in size, though most are approximately one hundred feet long, fifteen feet wide, and four feet deep. Depending on the size of the hatchery, there may be dozens or even scores of raceways. Water (usually drawn from a nearby stream, although sometimes from underground) flows—rushes, to be more precise—into, through, and then out of the raceways. Screens prevent the young fish from escaping with the current.

A tour of a salmon hatchery leaves one with the unmistakable impression that it takes a lot of technologies to tool salmon. Thomas P. Hughes noted that technologies are themselves "embedded in technological systems" that "consist also of people and organizations";[16] the technologies and the organizations together comprise *socio-technological systems*. Melvin Kranzberg, an historian of technology, similarly observed, "Every technical innovation seems to require additional technical advances in order to make it fully effective."[17] An immense amount of electrical and human social energy is needed to operate most hatcheries in the Pacific Northwest, and the organizational and administrative coordination necessary to bring off the hatchery performance is impressive. Water must constantly flow through incubation trays, rearing tanks, and raceways (alternatives exist, however).[18] Food, specially formulated after a century of trial and error, must be shipped to hatcheries, most of which are in rural areas. The proper number of eggs for each eight-by-ten-inch incubation tray has to be counted. And biologists, technicians, managers, secretaries, tour guides, and janitors have to play their respective roles for the hatchery to operate successfully.

Once they are placed in the raceways, young fish are "hand reared"—fed by a technician who scoops "fish chow" from a bucket and flings it into the water—for several weeks until they learn to tap a metal rod hanging into the water beneath automatic feeders. The feeder drops food each time it is touched, making it constantly available to the fish. Hatchery salmon are densely packed in their raceways. The fish have no shading from the sun, there is no variability in the water flow around them, their food is delivered by humans or automatically, and there is netting over the raceways of some hatcheries to keep away predators. The fish become aggressive, racing toward food, and they bunch up into huge schools. But they grow fat, happy, and quickly, a model of rationality.

Upon release, hatchery salmon often are poured directly into streams and rivers that are home to wild salmon (as well as other fish species). Some biologists argue that the huge numbers of hatchery fish commingling with wild salmon make it difficult for the latter to survive, simply because they are outnumbered. Thus, an unintended consequence of the manipulation of hatchery fish has been a contribution to the long-running decline of wild salmon.

Because of the decimated numbers of wild salmon, hatcheries have gone to great lengths to identify "their fish" both for biological purposes and in the hope that wild fish caught by anglers will be released. All hatchery fish have the adipose fin (the small fin on the back near the tail) clipped off or are otherwise marked. In the name of science, politics, and economic expediency, the fish are maimed, although biologists say that the adipose fin is the salmonid equivalent of the human appendix—vestigial and of little practical value.

Efforts at curbing disease and reducing the reliance on energy- and labor-intensive operations by employing technologically sophisticated machinery have failed to a large degree. An example is Dworshak National Fish Hatchery in Idaho. Constructed in the late 1970s, Dworshak was supposed to constantly recirculate its water, "passing through"—releasing—only a small portion of the water in use at any one time. The water was to have been temperature-controlled, treated with ultraviolet light to sterilize it, and filtered through sand for further purification. A computer-controlled pneumatic device would have replaced human labor in feeding newly emerged fish in "nursery tanks." But none of those technologies worked properly. Today only a portion of the water is recirculated, and the water is only heated during the coldest winter months. The dreams of technological rationalization in salmon hatcheries were destroyed by a Nature that failed to cooperate with the best-laid schemes of engineers, fisheries bureaucrats, and salmon biologists—and the dominant version of the physical construction of salmon.

PRODUCTION IN SALMON BIOLOGY

Historically, salmon biologists have been trained to emphasize economic and recreational uses of the salmon over all else, and where they apply their craft—especially in salmon hatcheries—the rationalizing tenets of capitalist economics, such as efficiency and productivity, are the yardsticks of success. In so doing, salmon become commodified. A biologist working at Northwest Salmon Hatchery told me, "If you've got bad production coming out of a hatchery, the first ones that'll know it are those people down at the dam collecting those fish. They say, 'Yeah, that looks like a [smolt from another hatchery] because it doesn't have any fins. That looks like a Northwest. Jeez, you know Northwest looks pretty good this year. You don't see any fungus on them."

"Production" is a salmon biologist's term for the number of fish born in any place, whether it be a wild stream or a fish hatchery, and it always carries economic connotations. Because the concept of salmon production is universally applicable to wherever juvenile fish are found, it permits biologists to draw comparisons between wild salmon and hatchery salmon. One agency biologist said of Northwest Hatchery, "By and large I think you'd have a hard time arguing that it's not successful in terms of continuing that run of fish. In fact, there are some years when that's basically the reason for a good fishery down on the Northwest [River]. That often produces a lot of fish. It's not without its problems, but they've learned how to raise fish. Maybe we've changed the fish so that we could culture them."

Production connotes a concern for the *quantity* of salmon, and implicit in this quote is the possibility that hatchery production may exceed wild production. In practice that is the case throughout the region. By some estimates only one-tenth (250,000) or fewer of the adult salmon returning to spawn in the Pacific Northwest of the United States are wild salmon. A far greater percentage of the juvenile salmon migrating to the Pacific Ocean in any year are likely to have been raised in hatcheries than to have reared themselves in the wild—outside of the hatchery. But today biologists believe that these hatchery fish lack the skills necessary to survive in the wild, in the free-running (and not so free-running) rivers and streams where they are sent at the time of release. This means that a greater proportion of them die in the wild as smolts than those fish that were born in the wild. By these measures hatchery fish are inefficient, and therefore irrational, because so much money and effort is devoted to fish that fail to survive.

Nor is the production perspective limited to those whose work is as applied as that of a hatchery biologist. One fisheries professor said, "There is an awful lot at stake on the Fraser [River in Canada]. You've got sockeye runs that are capable of producing 100 million fish—adults. You've got a tourist industry that could be built around that that would probably commensurize the economic value of that of the fishery. There are just a lot of benefits of access up that wild river." The implication is that economics is the chief reason why Canadians should resist damming the main stem of the Fraser—the Fraser River itself, as opposed to its tributaries, some of which already have been dammed. Damming the Fraser would result in an economic loss that could not be recovered in hatcheries because of the hatcheries' inefficiency. Of course other, competing economic interests—especially electrical power generation—may overwhelm narrowly salmon-based economic arguments against dams.

Production has another distinct and significant meaning within salmon biology. Hatcheries and other research-oriented endeavors (understanding that hatcheries' primary concerns are with producing fish and that the research undertaken at hatcheries is a rational means to achieving that end) produce biological facts. Bruno Latour argued that facts are produced through social interactions. He wrote that "a fact is what is collectively stabilized from the midst of controversies when" the statements of earlier scientists are confirmed by the statements of later scientists.[19] Once established, wrote Stephan Fuchs, "Facts have no visible authors and appear timeless and universal. . . . Artifacts are constructed, facts are revealed. Statements are spoken by subjects, facts let reality speak for itself."[20] Of course, facts do have authors: authors who make mistakes, whose data may be reinterpreted by later authors, what have you. But for the time being the scientific community stamps authors' assertions as fact. Salmon are produced as new *social* facts when biologists exert control over both the fish and the process of fact creation.

HATCHERY POLITICS

Politically, Northwest Hatchery was required as mitigation for a dam that blocked the Northwest River to salmon migrating upstream. While federal law directed the construction of the hatchery because of the massive dam's impact, in turn the hatchery allowed the salmon to become part of a social-technological system serving institutional interests. Politically and economically powerful groups benefited by preserving the salmon, a uniquely large strain, for the recreational fishery on the Northwest River. The hatchery was promoted as ensuring a continued sport and tribal fishery in the region, hundreds of miles from the mouth of the Columbia River. The resulting annual benefits for the state's economy are estimated to be in the millions of dollars.

But in the region those political guarantees, and the resulting economic and cultural returns, were some time in coming. In the 1970s and early 1980s several hatcheries were sites of protests by Native American tribes who argued that they were entitled to fish even though the number of returning adults was considered below that necessary to reproduce the next generation. It was a time of intense conflict over salmon throughout the Pacific Northwest, with the State of Washington fighting with several tribes in court over Native American treaty rights. The guarantee of fish for all appears to be sound today, however. A Northwest Hatchery biologist told me,

> It's all behind us as far as having a tribal fishery and not sport fishing or you could catch one fish and then you would maybe be closed down or there'd be a conservation closure made. We went through that whole period of time here. The community was all up in air about . . . it was so visible to see the tribe could fish. . . . And those same things were going on down at [other hatcheries] with the . . . police down there and they were routing the traps and all that. But it's all turned around now because everyone is sitting down with the tribe.

The confrontations have subsided, including those in Washington that led to the "Boldt Decision," a court ruling promising that Native Americans had first rights to fish for salmon when the state said there were enough to allow fishing at all. Native Americans have established a network of hatcheries that rival those operated by state and federal agencies in size, number, and complexity, ensuring continued production of salmon for their use and profit.

Production-based conflicts persist between Native Americans and the government, however. For instance, in 1994 the Nez Perce tribe imposed a fishing license fee for nontribal members who sought to fish for steelhead trout (technically a salmon species) in waters that flowed through tribal lands.[21] Both contemporary and earlier conflicts between cultures have at their root differing cognitive constructions, although the fact that Native Americans increasingly rely on hatcheries to produce their fish indicates that today's conflicts are less between cultural entities (with their distinct and nonoverlapping constructions of the salmon) and more between political entities (nations asserting their "sovereign rights"). In a time of heightened self-identity among Native Americans and rediscovery of their cultural heritage, other conflicts may develop as well; several Canadian biologists commented on the potential impact of the nascent "aboriginal rights" movement in their country.

HATCHERY ECONOMICS

Salmon hatcheries are important economic entities as well as political objects. Any given salmon hatchery contributes to at least two "fisheries," perhaps more. One is in the ocean, where the salmon live for from one to five years before returning to their natal streams or hatcheries to spawn a new generation. The return home creates the second fishery, which is near the streams or hatcheries of their birth. Other fisheries may be located along the return route as well. A third class of fishery is those operated by Native American tribes for their exclusive exploitation in accordance with treaty rights; it was a conflict over who controlled the fisheries on Nez Perce land that was the object of the 1994 conflict mentioned above. Hatcheries are also important to the local economy because most hatchery employees live in small towns nearby, and the hatcheries also provide some seasonal work for townspeople. In 1998, the Northwest Power Planning Council and BPA spent about $50 million supporting hatcheries; states, provinces, tribes, and federal governments spent several times that.[22]

Salmon biologists play essential roles in the pursuit of the governmental/economic ends embodied in hatcheries; they are as much a part of the hatcheries as the incubation trays and raceways. Biologists' fisheries-oriented socialization makes many of them amenable to the highly rationalized, historically dominant (though presently evolving) hatchery concept: produce the most fish possible for the lowest amount of money in the least amount of time. Hatchery biologists have sought to ensure that hatchery conditions are highly controlled and are ideal for gestating salmon. For instance, water is kept at a "perfect," steady temperature devoid of the occasional deadly swings that occur in the wild, and it is as sterile as possible to reduce the incidence of diseases.

In an even more explicit example of a physical construction of salmon, biologists eliminate "weak" salmon by "shocking" them while they are still in the egg stage. Holding the incubation trays waist high, biologists pour the eggs into an empty plastic bucket placed on the floor. The eggs that do not survive turn white in a few days and are removed from the trays, each of which holds hundreds of eggs. What remains after shocking is the cream of the crop, and few of these salmon die as they mature. This increases the economic efficiency of the hatchery, thereby reducing the cost of rearing fish because no food, labor, or energy is wasted on fry that cannot survive the rigors of hatchery life.

As further evidence of the quality of the hatchery product as measured by its ability to reduce inefficiencies found in the wild, biologists note that the oceangoing steelhead from some hatcheries are turned out on a yearly basis, although in the wild they rear for two years before going to sea. This is achieved by increasing water temperatures and available food, and it effectively doubles the hatchery's production, since the fish only have to be housed and fed for one year before they are turned out.

The outcome is a highly engineered fish. I use the singular case deliberately, for the historical goal in hatcheries has been to create a homogenized salmon. Ideally the individual fish behave the same. They emerge from their eggs at the same moment and mature at the same pace; they may be released at the same time and will return to the hatchery as adults in a narrow "window." Such fish are supreme examples of rationalization. They are products of a predictable, calculable, efficient, productive process—fast food with fins.

CERTAINTY, PREDICTION, AND TOOLING

Often, tools are created with built-in variability so that they can be manipulated to suit the tool user's differing needs; an example of such a tool is the adjustable wrench. Another is hatchery salmon. The direct users of the salmon tool are biologists working at the behest of politically-oriented governmental agencies. In the early years of Northwest Hatchery's operations, biologists obtained the eggs needed to produce the next generation by breeding the first fish to return to the hatchery. As soon as enough fish were mated to meet the hatchery's requirements, breeding ended. One agency biologist with a long familiarity with hatcheries told me,

> The run is normally bell shaped, but they got right up to the top of the bell and stopped. They had all the fish they needed and they stopped. They didn't take any from the second half of the run. That's a direct selection, and that by and large has stopped. We've been successful in not letting them do anything like that. The other selection is inadvertent in two ways: the earliest spawning fish get to rear the longest in the hatchery, and they're the biggest. And the biggest fish usually survive better than the smaller fish. So the longer you're in operation, the fish that are going to keep coming back to you are the offspring of the early-spawning fish just because they get bigger.

A similar story was related in Tim Armstrong's quote at the beginning of this chapter. The effect of this and other physical constructions of the salmon is to force the fish to work as biologists wish them to. A biological factory, the entire hatchery complex is tuned like a machine, a manufacturing artifact made up of fish, nonbiological technologies, bureaucracy, biologists, managers, and others.

Frequently, though, salmon behave like anything other than a reliable instrument. They cause "problems," although ultimately it is society that causes the problems through our continued manipulations of the fish. Once tooled and brought beneath the mantle of human control in hatcheries, scientists say salmon change in fundamental ways. Like other tools, salmon lose strength through time—in this case the ability to survive from generation to generation, or "fitness" in Darwinian terms—as a consequence of the process of being remade. This makes biologists anxious. Tim Armstrong told me, "That's what we're looking at now is ways to bring our hatchery fish closer to what that wild fish is. I don't think we'll ever be able to get them equal. You just can't duplicate wild. But you can get it closer than we've got." Little about the salmon appears certain other than their variability, whether the cause is genetic or environmental—including human-induced changes.

Certainty, then, is an impossibility with hatchery salmon. The next best thing, prediction within a range of error, is easier to achieve. It is "known"—that is, it can be predicted to fine tolerances—how long eggs will take to hatch, when the hatchlings can be moved from egg trays to rearing tanks and then to raceways, and how large a fish born on a given day will grow before its release from a hatchery. Prediction is important because the hatchery is not unlike a gasoline refinery. It is imperative that the raw material (crude oil or salmon eggs) for the final product (gasoline or juvenile salmon) flow through the plant (the refinery or hatchery) according to a precise schedule. Along the way additives (engine cleansing agents for gasoline or the food, hormones, medicines, and substances administered to the fish for scientific purposes) must be given at the proper time and in the correct proportions. No delay is tolerated. Slowdowns in a manufacturing facility's production may spell economic doom.

Similarly, if the timing for releasing young fish into rivers is off by even a week, the altered production schedule may prove politically and economically disastrous. For instance, a hatchery biologist who wanted to release chinook salmon two weeks later in the year to mimic new data on their wild cousins recounted a conversation with the hatchery manager. The biologist told me that "to change the time of release you have to start with the egg. It's hard to conceive, but we have to go back [that far] because we've got two year classes in lockstep behind each other. I said, 'Okay, I want you to hold these this year until April 15.' He says, 'I can't do it. I've got fish in the next year class right behind these. They have to go into those same containers in those same raceways, either that or they're going to die in here.' " Precision scheduling is in place, made possible only because the fish have become so predictable thanks to the hatchery's physical constructions of them.

However, even prediction has its limits. Complex, interacting systems—in this case, social systems, engineered systems, and biological systems—may render prediction unstable at best. Geoffrey M. Hodgson explained that in such complex systems

> tiny changes in crucial parameters can lead to dramatic consequences. The result is not simply to make prediction difficult or impossible; there are serious implications for the notion of reductive explanation in science. We cannot in complete confidence associate a given outcome with a given set of initial conditions, because we can never be sure that the computations traced out from those initial conditions are precise enough, and that the initial conditions themselves have been defined with sufficient precision. Hence in chaos theory the very notion of explanation of a phenomenon by reference to a system and its initial conditions is challenged.[23]

The systems perspective so central to the engineering ideal cannot account for the uncertainty embedded in those very systems.

AN AGRARIAN MODEL FOR FISHERIES

What could be more characteristic of our age than a modern factory, whether it produces gasoline or fish? In a time when "nimble manufacturing" and "just-in-time" delivery are the bywords for a lightning-paced, highly competitive economy that constantly aims to increase efficiency and meet immediate consumer demand, a two-year production delay like that confronting the biologist above seems quaint, if not anachronistic. Yet this agrarian model of production pervades hatchery biology at the same moment when industrialism reigns supreme, and the salmon are caught in-between. As living organisms, salmon have their limits. Although their gestation can be cut by one-half to one-third, fish scheduled for release cannot be held aside in silos for a time while a new production schedule is put in place. Once on the schedule, they must move through the system apace until their release.

So perhaps it is ironic that for hatchery biologists the agrarian model implicit in the domestication of salmon is the ideal, not the industrial factory model. Sam Freiberg, a federal agency biologist who has worked on hatchery issues for years, observed,

> With all the diseases and everything else [hatchery operations are difficult to undertake], yeah. And so you get a lot of those kind of

> things going on and it just takes time. No reason that fish would be any different than any other kind of agriculture. It takes some time to learn how to do it well. And that's part of the thing. The trouble with raising anadromous fish is that we've got to send them out to the ocean where they have to survive. They've got to survive the trip down and they've got to survive while they're out there. It's a really rigorous test compared to planting some crops and stuff like that, where you can keep your eye on it and tend it. So it's a more difficult situation.

The agrarian metaphor is far-reaching in salmon biology. For instance, it was said that one hatchery "sows fish up in every part of the . . . basin"; the salmon are "seeds," and often there is talk of "underseeded streams" for which it is judged that "carrying capacity" is adequate to support more salmon than are found there. Carrying capacity is a term that originated in range sciences to indicate the number of livestock that a given area can support.[24] In salmon biology carrying capacity is more elusive and imprecise; "supplementation" efforts to increase the density of fish in a stream—and thus more efficiently use the carrying capacity there—by sowing hatchery fish often fail. Similarly, the high seas "range" where salmon are said to "ranch" themselves may have its limits, as one biologist noted when he said, "Some people argue that, yes, we can put out too many coho salmon off the coast of Oregon and Washington. It's less compelling, I think, for chinook and steelhead, especially those that migrate a long distance up in the Gulf of Alaska." For some species even the vast Pacific Ocean may not be able to support the numbers of salmon being sent there.

The agrarian metaphor can also be found when hatchery and nonhatchery biologists refer to "stocks" of salmon—those that migrate to a certain point. And in keeping with the livestock submetaphor, one biologist told me that many hatcheries "brand the fish like they do a cow, right on the side. It's kind of a black mark, kind of iridescent. You can see it. We know which fish are tagged by their [adipose fin] clip, so we look for a brand and we can tell what's going on," that is, from which hatchery the fish came, because the brands, like those of ranching outfits, are unique from hatchery to hatchery. Although not very popular in the region, an altered form of livestocking salmon, "fish farming," is practiced as a form of aquaculture around the world. Any type of intensively human-controlled raising of salmon is referred to as "fish culture."[25]

And the agrarian model goes even farther. Salmon are artificially bred—that is, they are bred by humans under controlled conditions—just like many cattle; the unfit are eliminated and the rest are administered hormones and moved as necessary while in the hatchery. They eat at will—almost as in a feedlot, although this aspect of life history timing is opposite the fattening-up period for cattle because it occurs early in the life of the fish, not near slaughter as is the case for cattle. Once released, the salmon roam the ocean's wide open spaces, grazing on microorganisms and later on other fish. While in the ocean, salmon are "harvested." One author noted, "The U.S.A. produces close to 40 percent of the world's salmon. . . . Alaskan production accounts for 80 to 90 percent of U.S. harvests."[26] Salmon are a big money crop, and all that is missing from this scenario is a seagoing cowpoke. No range riders are needed, however. Salmon are anadromous: they have a genetic predisposition to return to the site of their birth.

HATCHERY SALMON AS "DIFFERENT"

As pervasive as hatchery-bred salmon are, one could be forgiven for assuming that biologists unabashedly support fish culture. While it is correct that nearly all biologists see a place for hatchery salmon, nearly all of those "enthusiasts," as they were known a century ago, object to how salmon hatcheries do what they do: the particulars of the ways in which the agrarian model is applied. Norm Wilkes, a fisheries school professor and ardent defender of hatcheries, calls the objections "hatchery bashing." Sitting in a carpeted and neatly kept office with copies of books he has authored and edited on the shelves, he classified the primary objections:

> One of the main ones is that hatcheries have changed the genetic integrity of the stocks. Introducing fish from outside of the watershed is part of it because you've introduced new genes. Because when fish come back, you don't get just the fish that you released back. And they don't all come back to you. They go over and mix with the wild population. So if it's a so-called exotic, or at least non-native to that

> habitat, you're influencing the genetic diversity within that population by overplanting with fish that are outside of that group. That's been a major issue. Selective breeding in the hatchery, at least the way it's been done in the past, is leading to an inferior fish, one that survives very well in the hatchery, but doesn't do well in nature.

There are actually three key objections in Wilkes' comment. Two are tied to genetics, and the first of these holds that the formerly common practice of "implanting" salmon from other areas into a new watershed is inherently ruinous. This is an example of the interchangeability of salmon concept introduced in Chapter Four. A Canadian biologist explained, "In the Fraser River populations that migrate one hundred miles have a different program, genetically, than populations that migrate eight hundred miles. We found that out the hard way. When you want to transplant the one hundred mile to the eight hundred mile, or vice versa, the one hundred mile transplanted to the eight hundred mile stops migrating after one hundred miles. The eight hundred mile transplanted to the one hundred mile keeps on migrating, so it's disastrous."

Genetic research comparing wild—not hatchery—sockeye salmon and their landlocked cousins, the kokanee, appears to support this observation. For sockeye and kokanee that are found in the same lakes, and those lakes are known to have been implanted with "exotic" kokanee from distant parts of the region, biologists say their evidence indicates that few of the implanted fish survive for long. Their explanation is that the implanted, exotic kokanee lack the genetic predisposition—programming, in the biologists' mechanistic language—to survive in dramatically different conditions. DNA analysis indicates that kokanee and sockeye populations thought to have evolved in the same lake are more closely related than would be expected had exotic genes been retained in the kokanee population. The same concept is said to hold for hatchery populations as well.

The second major objection to hatchery salmon based upon genetics emphasizes the prospect of hatchery salmon interbreeding with wild salmon. Biologists are uncertain how widespread this phenomenon is, but the possibility of it has given rise to whole new classes of salmon. Formerly, a distinction was made between hatchery and wild fish. More recently a third class of salmon has been created, one of ambiguous parentage: fish neither of entirely hatchery nor wild origin. This is the "natural" salmon. A hatchery biologist explained, "We call it 'natural spawning.' There's a terminology that is starting to get uniform. Natural spawning is where you have hatchery influence and wild spawning is where there has been little or no hatchery influence. Of course, this [hatchery] has . . . influence here, so it would be what we would consider natural spawning. . . . How successful it is, we aren't sure." In other words, some "hatchery" salmon, those bred by biologists, return to spawn the next generation in the streams and rivers *near* hatcheries but not *in* the hatcheries as biologists intend them to do. They are unintentionally beyond the control of the hatchery operators, like the breeding dinosaurs in *Jurassic Park.* And to hear biologists speak of it, the consequences are nearly as disastrous. The gene pool of the wild salmon is thought to be "polluted" when hatchery and wild salmon mate. Recent research indicates that hatcheries may need periodic infusions of genes from wild fish in order to avoid genetic problems, but if hatchery and wild fish have been mating, the truly wild genes my be difficult or impossible to locate. Hatcheries may inadvertently doom themselves.

This inferiority of hatchery fish is the subject of the third objection implied by Wilkes: raising salmon in a hatchery is fundamentally a waste of time. Biologists argue that hatchery fish are inferior to wild ones because they lack the skills necessary to survive in the wild, where they must live most of their lives after their release from the hatchery. This assertion is actually more behavioral than genetic: hatchery salmon have not *learned* how to survive in the wild. Behavior appropriate to the hatchery—for instance, racing toward a figure that looks like a human feeder and staying near the surface of the water to get at food quickly—may be deadly in the wild, where processes of natural selection eliminate young fish that fail to learn that safety in daylight hours is at the bottom of the stream.[27] A hatchery biologist addressed these concerns in Darwinian-Skinnerian mechanistic terms, saying, "We know there's a natural selection process. If you're asking why does a wild smolt survive better, yeah, I think the wild smolt, by the time it moves downstream, has been subjected to all these selective factors. It has been conditioned on the predators, it has been conditioned where to be and where not to be, the groups of fish that it's moving with are probably not near the size, the social and the stress size interaction with the hatchery fish. So it's a different critter." In the end, hatchery salmon die in large numbers once they are released into streams and rivers, and this counterbalances the enormous production of the hatchery system. The efficiencies end at the pipe where the juvenile salmon are poured into the streams. All that tooling of the fish is for naught when they fail to flee from predators or cannot find food. The hatchery's physical constructions have serious limitations.

Other objections to hatcheries exist as well. However, in contrast to the foregoing examples, these tend to emphasize a direct human role in decimating wild salmon rather than to stress genetic or behavioral characteristics of the fish. Thus the fourth objection: the problem of mixed-stock fisheries. In the words of a state agency biologist, a mixed-stock fishery

> creates a high exploitation on the hatchery run. It focuses the fishery on the hatchery run because they don't want many fish to escape to a hatchery; you don't need many fish to get the egg take. So the hatchery is there to provide a fishery. The fishery focuses in on that run, and in the process of doing so it overfishes the natural runs in the vicinity. This has become a coastwide phenomenon. Hatchery fish travel out of the Columbia and Puget Sound and northward to Alaska waters to feed; when returning as adults to these different river systems, they're fished all along the way at a higher exploitation rate than wild fish can withstand, and they greatly deplete the natural populations in the process.

Because hatchery and wild fish swim together on the high seas and are the objects of intensive fishing while in the ocean together, an unintended consequence of hatchery salmon is that they speed the depletion of wild salmon populations. Distinguishing between hatchery and wild fish is difficult at best when the fish are being caught by the ton; indeed, because some hatchery salmon are not marked for identification, and because seagoing fishing boats catch huge numbers of salmon at a time, making such distinctions is impossible. Salmon are thus tooled to death ultimately because of the inability of fishing technology and the fishing industry to catch only hatchery fish.

Tom Papas, an agency biologist who often works with hatchery personnel, identified a fifth and final objection to hatcheries. He said,

> Another glitzy thing is these doggoned hatcheries. They definitely have a place, but you get these jokers running around who have a

> tool and they want to use it everywhere for everything. "Captive broodstock is going to save the day for anadromous fish populations" or "Supplementation's going to save the day." In some cases I would recommend that we initiate those programs. Maybe lots of them. But don't let the cart get before the horse. Don't lose perspective. In many cases that's the least important thing you can be doing. Let's get out there spending that money buying habitat or buying some water, which is buying habitat. Mainly it's so damned frustrating seeing this glitzy stuff run away with the show.
>
> *Q. Glitzy inherently means high cost?*
>
> PAPAS. Technology and usually high cost. We're so addicted to and so expectant that technology's going to save us. I'm obviously biased. I'm from the school where I really like the idea of sustainable agriculture. To a large extent it's going back to the way my grandpa used to farm. Don't rely on technology for all the input. You rotate with clover to get some of your fertility back. We're dealing with these ecological systems, and I think this stuff is highly appropriate and practical.

Those "jokers" so eager to tool salmon have politics and economics on their side. As unsustainable as the hatchery form of agriculture may be, Papas recognized in it a political and economic force. It puts fish into rivers and the ocean where anglers and commercial fishers have a chance to recreate and to make profits.

There is a substantial amount of inertia to removing salmon hatcheries, because they are politically popular and economically valuable. Commenting on hatcheries' role in his government's enhancement program, a Canadian biologist said, "So this is a very obvious conflict between the enhancement people and management people. But the integration with our fisheries managers was really poor in the beginning of the program. Of course, now you're left another kind of legacy of these things sitting around. You can't turn them off. You've got too much social pressure. What we're trying to do now is look at how to optimize production by experimenting with them. Even that we may have trouble getting support for."

Economically, Papas's dream of buying habitat instead of hatcheries seems far-fetched, for the substantial cost of water or streamside land would be difficult to bear in times of governmental belt-tightening. And it is doubtful that taking energy- and labor-intensive hatcheries out of production and spending that money on habitat would yield as many salmon, thereby harming and not enhancing the productivity so important to rationalized systems.

THE PRO-HATCHERY RESPONSE

Hatchery biologists have an understandable attachment to their places of employment. But those who manage the bureaucracies that implement hatchery programs in Canada, in particular, have much more at stake in hatcheries' continued existence. In the 1970s extravagant claims were made regarding the benefits of hatcheries. At that time some of Canada's most respected salmon biologists expressed strong opposition to British Columbia's enhancement program. Although the enhancement concept includes numerous approaches to improving salmon production both in hatcheries and in the wild, including habitat improvement, the biologists' objections centered on a C$700 million hatchery construction plan. A split developed between government salmon scientists and others in the federal government, and the rift was reified—made concrete—in the form of separate subagencies for enhancement and for scientific research.

This is a classic instance of the rationalization of Nature through bureaucratization. Twenty years ago hatchery advocates argued that hatcheries were a "proven technology." "Now," says a Canadian biologist, "they're admitting that, 'Well, we said things like that because we had to get the money first. We didn't really believe that.' But there's an unfortunate problem, because when you do get the money and then you want to study it, you're told, 'Why do you need to study it now if you told us you didn't need to know in the beginning?' " Years later, social, political, and economic inertia seem to have ensured the continued existence of the Canadian hatchery program. Furthermore, the studies that scientists say they need to conduct to assess the failings of this "proven technology" cannot be funded because of the technologically optimistic rhetoric of yore.

In the United States, however, hatchery advocates have begun to answer their critics, first with frank admissions that the ways that they have physically constructed the salmon may be problematic and, second, with self-assurance that they can create salmon that address at least three of the five objections above. Tim Armstrong put it this way:

> Well, we're going through a changing period. It's the biggest change that we've gone through in looking at our hatcheries and how we're going to manage them and how they integrate into the overall picture of fish production. We've been using hatchery production in the basin since the late 1800s, at least down in Oregon. I think it was over a hundred years ago that the first hatchery was built. These up here, of course, for salmon and steelhead, have been primarily since the late sixties and on up through the present day. It seems like in the last ten years hatcheries have been completed every year, every other year. . . . So we're still building hatcheries in an effort to solve our problems, and I'm not so sure it's going to solve everybody's problem. We're having to take a different look at our hatchery and say, "How can we do a better job of producing the fish that we want to?" I think that sometimes we have sent the wrong message to other groups, saying that hatchery fish are great and they'll replace a wild fish or all the rest.

The root problem with hatcheries, according to this line of argument, lies neither in failed technologies nor in a lack of understanding of salmon genetics. Rather, it is the failure of the bureaucratization and rationalization of the hatchery process. More specifically, the issue is said to be "a people problem." One biologist summarized this when describing the joint (and now defunct) U.S.-Canadian Salmon Commission's approach to hatcheries: "It's not really the hatchery, it's how we manage them. If we do stupid things, we can't expect good results."

Hatchery advocates tenaciously argue that scientifically rational, effective, and environmentally benign hatchery management is possible, given enough knowledge. That knowledge did not exist years ago, argued hatchery advocate Norm Wilkes:

> So if you go back to the thirties, the Mitchell Act said, "Okay, we know fish have come up here to spawn and they're not going to do that any more because we've got these dams up here. So we're going to build a hatchery system to produce fish and mitigate against the dams." Conceptually, that made sense. As long as you put fish back, what the heck?

> We understand now that that may not have been a smart move. But given the state of the science in those days, it was not a stupid idea. It made sense to them, and it probably made sense for a long time.

The same was true in Canada. Today, however, Canadians are seeking the necessary knowledge to create environmentally friendly hatcheries. According to one biologist, this effort is proceeding through "large-scale experiments. If you really think that there's marine carrying capacity out there, there's also a lot of environmental variables or variation, so you've got to make major changes in the system before you detect them. What we're talking about is literally taking a suite of hatcheries and turning some of them on and off, or turning them way down and others way up, this sort of thing, over a fifteen- to twenty-year study. It would take that long to actually determine what's going on."

In the United States, studies are underway to examine the effects of supplementation. This means "assisting" the salmon—increasing production in the wild—without harming wild fish. Like the proposed Canadian hatchery experiments, studies in the United States on supplementation are long-term. "Can we use hatchery fish successfully and safely with wild fish?" one interviewee asked. "It's going to take twelve or fifteen years to get answers." What is fascinating about this is that decades ago hatcheries were portrayed as quick and easy methods for physically constructing salmon predictably. Now it appears that only through deliberate, long-term research will biologists be able to determine if hatcheries have a role in an uncertain future for the salmon.

More immediate efforts to retool hatcheries, and thereby to physically reconstruct the salmon they produce to more closely mimic the characteristics of wild salmon, are being undertaken as well. A state agency biologist said,

> . . . I don't think we've fully explored how we can best utilize cultured fish. One of the things we're doing right now is attempting to modify drastically how fish are reared under cultured situations. We're providing them with real gravel substrates in raceways, in-water structure, providing them with cover over the top of where they're going to be raised. We're going to be liberating food underwater at random during day and night. We're attempting to essen-

> tially create animals which have appropriate behavioral attributes as well as appropriate body colors. . . .
>
> In some respects perhaps we culture our fish to make our life as fish culturists a little bit easier. Of course pathologists and others say, "We've got to rear fish this way because if we don't there are going to be disease episodes" and so forth. I've always felt, philosophically, that we need, again, to have the fish tell us how we should be handling them and taking care of them in culture settings. Again, we need to ask them the right kinds of questions. They can tell us how they should be cultured.

Comments like these imply that the agrarian, mechanical, rational cognitive construction will continue to predominate in salmon biology even as the products produced by biologists are changed. The salmon have been given a voice, one that speaks in utilitarian tones. "Listening" to those products may be the answer befuddled hatchery biologists have been seeking for more than a century. Regardless, the tooling of salmon persists. Nearly all of the biologists with whom I spoke see salmon hatcheries not only as having caused problems but as playing a crucial role in resolving those and other problems. The very idea that hatcheries are problematic represents a changed meaning and a changed reality for salmon among the biologists who study them.

Constructing hatcheries and hatchery salmon as problems meant that an alternative, nonproblematical salmon had to be constructed within the profession. That salmon is the wild, untooled fish. Wild salmon have been declared the benchmark, the Weberian ideal type, for hatchery biology. Historically, there was little concern among hatchery biologists for maintaining the characteristics of wild salmon. The emphasis was on the quantity of salmon produced and not the quality of those fish. As mentioned, today's salmon biologists identify at least three reproductive classes of salmon: wild, natural, and hatchery. Those biologists, including both advocates and opponents of salmon hatcheries, accept the characteristics of wild salmon as *the* measure of quality. They also argue that more knowledge is necessary in order to physically construct hatchery salmon in ways that approximate their wild cousins. The vast majority appear to believe that such tooling is possible, given time.

NEW TOOLS FOR TOOLING SALMON: HIGH-TECH FISH

Salmon hatcheries are one of the most visible means by which the physical construction of salmon takes place, and they are where society's cognitive models of salmon are most easily seen. However, hatcheries are not the only examples of this concept. The dams that block nearly every major river in the Pacific Northwest are another example of how society physically constructs salmon. Extensive "bypass systems" at the dams are used to direct young salmon to barges. These "salmon taxis" shuttle juveniles below the last dam on the Columbia, where they are released to make their way to the sea. Barging salmon is said to be the most effective way of getting the fish through the huge lakelike reservoirs behind the dams. The water in the reservoirs moves so slowly that fish have difficulty finding their way downstream. In addition, the water may become too warm for the fish to survive in, and the reservoirs are home to predators. Still other technologies tool the fish, such as the fishing gear used for both inland and maritime fishing. And an array of examples of cognitive constructions may be found, for example, in the laws that mandate hatcheries, forms of fishing gear, and fishing seasons, as well as in the customs that direct Native Americans' ritualistic uses of the fish.

Two relatively new forms of constructing salmon bear special elaboration. One is the increasing use of salmon for in vivo research, and the other is the genetic alteration of salmon. Both of these use sophisticated technologies for tooling the fish. And like all examples of physical constructions, cognitive constructions are also found in these technologies.

WHEN SALMON RESEARCH THEMSELVES

Historically, salmon have been used to research themselves in many ways. For decades a variety of "tags," pieces of plastic or metal usually fitted onto or through a fish's fin, have been used on adult salmon to allow researchers to gather information about them. Anglers are sometimes promised gifts, such as baseball caps, for returning the tags to research centers with information about where and when the fish were caught. Biologists also "bar code" young hatchery salmon with chemicals that stain their otoliths, the bones in their ears. When the bones are later removed from the fish, information about their age and growth may be obtained by examining them under a microscope. Similarly, equipment such as underwater television cameras and hydroacoustic locators like those used by anglers to find fish underwater, have been used to observe the subsurface movements of salmon.[28]

The latest technologization of the salmon is an intensification of the tool-making process made possible by industrial processes known as "microminiaturization": the creation of new, very small technologies. Such processes are integral to redefining society's relationship to the salmon and thereby giving new meanings to the fish, and many salmon biologists are eager to take advantage of these technologies. Microminiaturization represents a refinement of the instruments of control. It allows the scale of the information-gathering technologies used to physically construct salmon to shrink, such as shifting from huge hatcheries, dam bypass systems, and barges to pinhead-sized transmitters.

Perhaps the simplest of these new research technologies is the "coded wire tag." Almost too small to be seen by the naked eye, these tiny pieces of metal are injected into the heads of young fish before they are released from hatcheries (to my knowledge, coded wire tags are used exclusively with hatchery salmon). Each wire is notched, and the microscopic indentations may carry information such as the location of the hatchery, date of release, and the like. When the fish return to the hatchery as adults, the tags are located using metal detectors, cut out, and read under a microscope. This allows researchers to track high-seas migrations and to conduct a range of other studies.

A more technologically advanced example of microminiaturization is the Passive Integrated Transponder, or "PIT tag." Pam Mitchell, a biologist who works on the Snake River, said that much of the research on young wild salmon relies on PIT tags. The first step in using them "requires a fairly large hypodermic needle that injects the tag into their abdominal cavity," Mitchell said. PIT tags are carried by the fish in their stomachs, providing valuable information to biologists because the technology "allows us to track an individual fish over its life cycle, given certain constraints on how that tag can be read. That's been quite a breakthrough," Mitchell told me. The microchip-like transponders are widely employed and have prompted biologists in the region to dub the salmon–PIT tag combination "fish and chips."[29] PIT tag information is automatically read by sensors at Snake River dams. This self-tracking by fish allows scientists to obtain detailed information on salmon migration.

Similarly, "radio tags" enable monitoring of adult salmon in an area such as a reservoir. Mitchell said that "our ability to track fish has greatly increased. When I first started years ago, you had ten frequencies you could use and the range at which you could figure out where a fish was was pretty limited." Today there are hundreds of separate frequencies available to scientists, allowing them to track many fish simultaneously, and the range is greater as well. Some of the radio tags may be quite large; one biologist reported forcing radio tags down salmons' stomachs. The fish were tracked thanks to transmitting antennas that went up the fish's gullets and trailed out of their mouths as they swam.

Even more technologically sophisticated means of tracking salmon are becoming available to biologists. A Canadian researcher described such advances:

> I suppose technology in itself has really progressed recently. I'm thinking of microprocessor-controlled samplers, for example. So you can now build these large sampling arrays for ocean studies, we're working on them now, where you set them out there and with a certain amount of power—maybe from solar or something else—they can sample at programmed intervals and collect certain samples. Hell, now we can transmit the information through satellite banks and get it back. We can track them by satellite to know where they are. There's no question that the technology has opened up what we're able to do.

These developments may allow salmon biologists to obtain detailed information about the most mysterious aspect of salmonid life history: ocean behavior. Moreover, the information may be obtained in an instant and without the biologist ever leaving her or his office (talk about efficient!). These technologies have important implications for the way salmon research is conducted and, with the use of less expensive yet more powerful computers, for the amount of information that can be gathered, processed, and interpreted.

These technologies demonstrate that, more than ever before, *the salmon have become research instruments in themselves*. No longer do biologists have to examine the fish for brands to determine their hatchery of origin, look in a book or make a phone call to determine when the fish were released from a hatchery, or work with small numbers of subjects for their migration studies. Thanks to these technologies, today it is within biologists' abilities to obtain extensive and readily manipulable data on every hatchery fish and on as many wild salmon as supervising agencies will allow. (All that is lacking is the funding for such an operation, although sophisticated statistical sampling regimes probably reduce the value of and the need for universal tagging.)

These television, tag, and transmitter technologies turn salmon into water-borne versions of caribou or wolves that commonly are fitted with collars so that they may be tracked and their lives intensively analyzed. (One difference is that most of these technologies are *inside* the salmon; they cannot be rubbed off, nor will they ever drop away, as is the case with many of the tags or collars used with other animals.) In effect, like many other species the salmon take the laboratory with them into the wild, as Bruno Latour argues Pasteur did in his studies of microbes.[30] Salmon *are* the laboratory. The technologization of salmon is increasingly complete—it now can extend to every movement, even in the deepest oceans. That technologization allows ever more rationalization, too. Calculability, predictability, and efficiency are all improved with better data and better data-gathering technologies. These enable society to have more control than ever over the salmon.

THE SOCIAL CONTEXT

This rapid rationalization of the research process is leading to changed meanings of salmon, not only for biologists but for society more generally. Although stiff competition for research funding is a given in the salmon biology community, governments at the national and state/provincial levels provide many millions of dollars to biologists using these technologies, much as governments did for hatcheries ten, twenty, and even one hundred years ago. The funding is available thanks to two factors. First, for more than a century political and economic entities have identified salmon as a "resource" deserving of this level of attention, and some recent legislation, such as the Endangered Species Act, has left government with little choice but to support research. However, salmon were intensively studied for decades, long before they became a "social problem."

Second, political and economic entities selected science as the arbiter of all questions regarding the salmon, and increasingly they are deferring to scientists' judgment. In some instances scientists *are* the policy makers. Marc Raymond was the lone corporate salmon biologist with whom I spoke, and his job is both research-oriented and has him working closely with regulatory agencies. He remarked:

> For a long time the economics drove policy, I think is the best way to put it. Decisions were based primarily on economic realities, whether it was economics related to forest practices or economics related to fish harvest. Our natural resource policy decisions were based on economics. That's changed dramatically here in the last five to seven years, and the Endangered Species Act has played a large role in that, certainly, but also the increased awareness of folks about what was happening to a lot of these resources has driven that as well. Now I think that science plays much more of a role in driving policy decisions. Clearly, economics is still figured into the mix and may still be the predominant determinant of the ultimate policy decision. But certainly science has become more of an important player in making these policy decisions. The appointment of Jack Ward Thomas [a biologist] as the head of the Forest Service is a very good indication of that. But I think what has also happened is we've put a lot of people in very powerful positions, scientists, who haven't necessarily been in that position before.

Placing scientists in decision-making positions indicates the pervasiveness of biology in all matters salmon. However, governmental and economic entities continue to exert control over salmon biology by effectively limiting the questions that can be researched by scientists.

Bruno Latour notes that "scientists and engineers speak in the name of new allies that they have shaped and enrolled,"[31] in this case the salmon. Scientists may assume this role by being anointed as spokespersons by powerful political and social interests as well. Canneries, commercial fishing fleets, dam builders and operators, barge owners, environmental groups, and government agencies: these are biologists' allies, too. These interests shape and enroll the *biologists,* who in turn shape and enroll the salmon. Increasingly, the salmon are the tools of salmon biologists, and this further empowers biologists' assertions and those of their social allies.

Something like this has happened before, when cannery owners along the Columbia River claimed all of the fish for themselves at the turn of the century.[32] But cannery operators seldom attempted to remake the salmon[33] and they had few if any biologists working for them. In contrast, salmon biologists today do physically construct salmon in order to pursue science's ends, which are, because of the dominance of structural entities, largely political/governmental and corporate ends. Aside from the salmon biologists, many other potential users of the salmon have been eliminated. For instance, in 1994 almost all high-seas fishing for Pacific Northwest salmon was banned, commercial and recreational, and fresh-water fishing seasons were severely limited as well, even for Native American tribes. Such regulations are usually temporary, as was the case here, but in the resulting void it was almost by default that salmon biologists were one of the few social groups with any more than an inconsequential claim to the physical use of the salmon.

GENETICS AND THE NEW SALMON

As advanced and impressive as many of the new tracking and analytic technologies may be to salmon biologists, many researchers are even more enthusiastic about the possibilities that salmon genetics may bring. Witness the excitement of George Williams, an eminent Canadian biologist, in describing the potential impact of genetics on salmon biology: "Molecular biology has provided tools for gaining understanding that we could never get before. You can speculate, well, for instance: You mentioned the Snake River sockeye. You lose it. Well, if I was an optimistic gene jockey I'd say, 'Don't worry about it. Just send me one of those kokanee and we'll fiddle around with a few of its genes and we'll get a gene from here and a gene from there and, bingo, we've got your new sockeye.' You may, if you're a doubting Thomas, say, 'Oh yeah? Come back in a hundred years.' The speed with which molecular biology is moving is absolutely incredible." So is the potential range of impacts. Here I will mention only two: the use of genetic alterations of salmon for economic and for conservation purposes.

One of the most frequently discussed aspects of genetics in the popular press is "gene mapping," the identification of the location and role of each of a given species' genes on its chromosomes.[34] Genetic maps of salmon have biological and economic implications, as Alex Stand, a salmon geneticist, explained:

> The trout and salmon are interesting to study in part because they have an interesting evolutionary history. They apparently went through a chromosome doubling at some point in the past, they think something like 50 million years ago. . . . It's something that happens a lot in plants, but it's thought to be less common in animals. So studying the aftereffects of that is an interesting research question that you can address in trout and salmon. Then the other area, of course, would be, "Can this mapping information provide us with some useful information for selective breeding and genetic improvement of farmed strains of trout and salmon?"

None of the seven species of Pacific salmon in North America have been genetically mapped, although a number of researchers are endeavoring to complete pieces of such maps. Once the genes are mapped, complete, widespread tooling of the genetic makeup of salmon will be possible.

For now, biologists have to be content with less detailed genetic manipulations, such as creating "triploids." A triploid organism has three sets of chromosomes rather than the usual two. This renders it sterile. Creating triploids has several advantages. For instance, fish farmers long have been interested in creating hybrids of different species of salmon to take advantage of the species' unique characteristics. In some cases their efforts have been effective, but often they have not. Stand commented that

> triploid hybrids actually survive, frequently, a lot better than a normal hybrid does. For example, you can cross a rainbow trout female with a coho salmon male, and normally none of the offspring will survive. But if you make a triploid, and we usually do that by heat treating the eggs at a specific temperature shortly after fertilization, you can make a fish that then survives that's two-thirds rainbow trout and one-third coho salmon. So it's interesting, then, to study their characteristics, see if they have some interesting traits. In fact, they found in that case, for example, that a hybrid is resistant to a virus called "IHN," which a normal rainbow trout is sensitive to. So it opens up a whole variety of research areas, I

> guess is what you'd say. It hasn't been tremendously widely picked up by the industry, but I still think there's quite a bit of potential for that kind of application.

He added, "Another aspect of this whole area of triploidy relates to the idea of sterilizing fish for conservation reasons. For example, there's been a lot of concern about hatchery fish escaping and interbreeding with wild fish and changing their genetic makeup. This definitely seems to be a big problem in some areas like Norway and Scotland where there are large salmon-farming industries. Some of us think that this idea of sterilizing the farm fish might be a nice compromise so that you could still have a farming industry and yet not have negative impacts on the wild salmon."

Genetics may have other roles in salmon conservation. For instance, "DNA fingerprinting" may allow for the quick identification of fish from differing stocks. Such techniques are similar to PIT tags but may be easier to use and more widely applicable, since it is not possible to PIT tag all salmon.

Stand asserted that DNA fingerprinting might have wide-ranging implications: "I think [it's] the thing of the future, mainly because of a lot of international conflicts and these fishery problems. We're going to need to do better at identifying stocks through, hopefully, DNA approaches as opposed to tagging and things like that that we've done in the past. I'm sure that's going to be around. There's a lot of need for that sort of thing." Biologists also see genetic techniques as valuable for answering conservation-related questions such as the differences that develop over time between wild and hatchery salmon. Some laboratories are storing "cryogenically preserved" (deeply frozen) salmon sperm with the aim of analyzing such questions fifteen to twenty years hence.

With genetics it appears that fisheries biologists may be able to both retain their long-term interest in producing economic returns to society while simultaneously addressing the cutting-edge concerns of the discipline, such as the preservation of specific runs, all through increased physical constructions of the salmon. Among the biologists with whom I spoke, there were no vociferous critics of genetic research (as there were of hatcheries). In our interviews, nearly all of them referred positively to the role of genetics in understanding salmon, if not to the genetic techniques that are now being used to obtain that knowledge. The most critical comment, by Tom Papas, was that DNA analysis leading to gene maps and genetic markers "is not that exceptional yet. It doesn't provide much more information than electrophoresis [a technique used for analyzing proteins that has been in use for decades]. But the bread-and-butter things, the things that are going to get us something in the [long] run, suffer time and again because the glitzy stuff is where the money is going to go. I think I see the same sort of thing happening with this DNA work. It's a shame that we get caught up in the fads so easily."

As with those who oppose hatcheries, opposition to genetic techniques appears to be pointless. Biologists familiar with funding mechanisms told me that grant proposals for genetic studies have a disproportionately large chance of receiving funding. As Bob Nicks, a professor of salmon biology, observed, "Basic biology of salmon: nobody's going to fund a basic biology sort of thing. It's not high tech. For example, understanding the life history and nature of salmon: we don't know, after 150 years, some of these things." The physical construction of salmon is being perpetuated through the highly politicized research-funding process, furthering the rationalization of the fish while ignoring basic biological issues. Meanwhile, the utilitarian construction of salmon proceeds unabated.

Conclusion: Salmon as Social Fact

Salmon hatcheries embody society's constructions of salmon through the multiple ways that they endeavor to tool the fish. Their existence alone is testament to the power of the economic and political cognitive construction—the economic and political meaning and reality—that society has imbued salmon with. The process of physically reconstructing salmon is advancing in other ways as well, and this, too, has implications for the imbued meanings of salmon. Hundreds of research scientists in the Pacific Northwest are using high technology instruments (including the salmon themselves) to examine salmon behavior, some of which has never before been accessible to science. Along with their peers studying the genetic blueprint of life, DNA, these scientists are closing some of the last remaining large gaps in industrial society's knowledge of the salmon. Once that knowledge is complete there still will be research questions to be asked about the salmon, some of which are fundamental, but the fish will be largely known to an industrial society bent on using salmon for its own ends.

The knowledge being created represents the possibility of a level of control over the salmon that only a few years ago was unimaginable. It will be known where to fish for salmon in the oceans at any time and how many of them there are to be fished for. Hatcheries will better mimic wild salmon, eliminating some of the major objections to hatcheries and perhaps eliminating any scientific basis for distinguishing between wild and hatchery fish. And geneticists may have it within their power to re-create salmon runs that are extinct or nearly so. The industrialized construction of salmon then will be complete. Salmon will have two identities. They will still be fish, but they also will have been reproduced by society as entirely new entities, new *social* facts.

Max Weber argued that economics, government, bureaucracies, technologies, and science together rationalize society.[35] These forces also rationalize salmon. The fish are brought into industrial processes and themselves become industrialized and technologized, not only by commercial and highly mechanized artifacts such as fish hatcheries, but through technologized and scientific processes such as tagging, tracking, and genetic analysis. To the extent that we seek to accommodate the salmon in our technological systems, they are industrial artifacts. Salmon, then, are already social-technological creations that are supported by, and that reinforce, society's most powerful interests—government and economics.

However absolute all of this may seem, the knowledge base upon which these creations rest is nevertheless contextual and constructed. That is, the knowledge is a social product, created by our times and our science. Don Ihde wrote that "the praxical 'knowledge' implied in technology points to an interpretation of nature itself."[36] In those technologies and their interpretation of Nature, we see a reflection of our society. The rationality macroconstruction that dominates nearly every facet of our lives shimmers in that mirror. We see salmon controlled through technologies based upon efficiency, predictability, and quantification. They are at risk of becoming little different than the products of hamburger joints, HMOs, or—increasingly—higher education: centers of mass production with little concern for quality, care, or conscience. Yet some biologists dare to question this dominant construction. We will explore their alternative in Chapter Six. First, however, we need to take a closer look at science and the stories it tells us.

5 Mythology and Biology

ONE OF CANADA'S most esteemed biologists, George Williams, struck me as a time traveler. He seemed to be living in two utterly distinct and incomprehensibly distant eras. One was a time when ambitious scholars read all of the literature in their field and were conversant with the current works in other disciplines as well. Back then, it was possible to move from academia to management to agency administration, and later to return to campus to take additional mathematics courses or even to teach. It was a time when reprints were the only economical means of obtaining other scholars' works and when positivistic science seemed perfectly tenable. Good science would lead to the discovery of natural laws. After that, control—never questioned as a norm—would be at hand, control in the hands of wise, trained, benevolent men (women would be in supporting roles, maybe even programming computers). The time was not so long ago—sixty, fifty, perhaps only forty years ago.

Williams also seemed to be living in today's hectic and scattered, complex and subdivided scientific world and reveling in it. No Reaganesque "those were the days" mentality for Williams. He has kept about him the positivistic, optimistic outlook of one who succeeded famously during a time of stunning transitions in his field and in the world around him, but he has no desire to go back to a past that never was. Reprints of scholarly publications? Don't be ridiculous: use compact disks with abstracts or even complete articles on them. Disciplinary focus? Intense specialization may be necessary now that knowledge is exploding, but isn't it amazing and exciting that knowledge is growing at such a pace? Question science? Sure, molecular biology might pose problems if bad genes are let loose, but my God think of the possibilities; there are legitimate concerns with nuclear energy, but France would be nowhere without it. Genetics will be to the world what nuclear power has long been to France: liberation!

Still active several years beyond retirement, Williams sat comfortably in a large office, attended to by his personal secretary and several graduate student assistants. He was jovial—his sense of humor is legend in the discipline—but always earnest and often critical. Out of all the things we discussed in our lengthy conversation, what seemed to trouble him most were indications that science was failing. Its promise and potential are too great to be wasted on the foolishness that is creeping into it.

One especially bothersome trend in science, according to Williams and others, is the "gray literature" produced by scientists in public agencies. Seldom if ever peer-reviewed, and by definition not published in reputable scientific journals, gray literature is a threat, Williams said.

> The failure to get things published, to put a lot of this stuff out internally without getting it critiqued by external reviewers, leads to the propagation of a lot of myths. It is so aggravating. You'll meet with a bunch of these people in government—this applies in Canada as well as in the United States—and you'll make some criticism of what they're doing in management. You'll say, "You shouldn't allow such a large catch," or something like that. Whatever. You make some criticism and say, "You know, there ought to be some study of that. Do some research to be sure you're doing it right." They'll say, "It's already been studied. We've got a report on that." You say, "Okay. Say, I'm curious. Can I have a copy of the report?" They give you the report and it's some pip-squeak exercise that you'd give a C-minus to an undergraduate, and this is the basis of policy. Uncritical analysis. One of my major beefs would be the failure to bring projects to completion, which is publishing in the literature where people can criticize it.

Gray literature represents failed control, the failure of agency scientists to submit to the will of the fisheries profession. Simultaneously, gray literature is achieved control to others. The same factors that exclude internal reports from peer review (primarily the closed, privileged character of bureaucratic structures) establish control over data and ideas *because* they are shielded from external review. In this process *myths* are made: the myth of depoliticized, objective science; the myth of science as monolithic and closed to interpretation; the myth of set, immutable standards that scientists know and dutifully follow. These myths play a large role in biologists' constructions of salmon, especially the dominant construction of the fish.

In labeling the concept explored in this chapter "mythology," I do not mean to cast aside all that science does or all that it has achieved as unfounded, baseless, and *merely* a set of beliefs. All realities are constructions, society's attempts to bring order to a chaos of beliefs and behaviors. Myths, too, are social constructions, meaning-imbuing interpretations of that which social groups identify as a reality. In using "myth" as synonymous with "social construction" I mean to challenge readers by using a familiar term in a way that is unfamiliar to many. Here, myths are not meant to be falsehoods or fantasies. They are simply stories portraying very real realities. But the connotations of the word myth are unavoidable, and that is a helpful thing for us. Science's stories are powerful and persuasive. But behind the scenes they can be messy: confusing, contradictory, poorly founded, uncertain, and in a sense impure. In order for any set of myths to maintain a position of power in society (for, as with all things powerful, they are cognitive yet may have influence on behavior) they must be true to their own ideals. As we shall see, that is easier said than done.

Science: Myth and the Material

Some constructivists would go so far as to assert that science is indeed mythical in the commonplace sense of the term: pure belief. They would not put it so boldly, preferring instead to speak of the knowledge that science produces as a "social product." But science-as-belief-system is what they are actually saying. Other scholars have taken umbrage at such statements, and it is appropriate to briefly consider that debate before we proceed.

Those who say that science knows Nature and that society has little to do with the creation of ecological (or chemical or physical) laws are known as "realists." They argue that society is bound by the same ecological laws as is *nature*.[1] Nature is not plastic, moldable to our wishes and whims, but is absolute. Further, they say, it is a mistake to argue, as some constructivists appear to do, that nature is merely the raw material for social endeavors. Doing so ignores the fact that, as Raymond

Murphy wrote, "the more humanity intentionally seeks to control the natural environment, the more unintended consequences it produces."[2] And perhaps worst of all, constructivists ignore the "immense dependency of humans on their natural environment."[3] The planetary biota is an interconnected whole, humans included, yet constructivists act as if society and ecology were as separate as different planets, two worlds without contact.

While the data presented here do not justify the extreme constructivist interpretation of Nature as belief and therefore as an artifact, neither is an extreme realist position tenable, either. Stephen Cole's observations in his book *Making Science: Between Nature and Society* point the way to a middle ground in this debate. Cole wrote that "it seems likely that constraint is imposed by the existing body of accepted knowledge in the same sense that positivists might say that constraint is imposed by nature. The accepted body of knowledge is the functional equivalent of nature."[4] Here he acknowledges one of constructivism's central assertions, that what modern societies consider Nature is largely dependent upon the knowledge base created by science. As such, Nature is the product of a social institution, implying the possibility that Nature is filtered through human action and is not directly, completely, or simply transmitted by science from a raw state into knowledge. However, Cole also wrote that "social variables cannot be used to explain why one model of DNA rather than another was accepted into the core. This position, then, although it rejects completely the rationalistic view of science, is a form of realism."[5]

Cole's realism is a tempered one that balances social and natural reality. His perspective comes down to this: Nature is fleeting, uncertain, and socially-created. Its reality is tenuous because science's understandings of Nature are subject to change as knowledge about Nature and natural entities accumulates. However, contra idealistic constructivism, science is predictive with a higher degree of accuracy than any other way of knowing available to us. And when ecological "laws" change, the new laws encompass not only old reality but address more recent anomalies as well.[6]

So here we treat Nature as something that scientists can give us insight into while we simultaneously acknowledge the influence that society has both on what scientists do and on the knowledge that they produce. Myths are the tenuous stories that biologists tell, the narratives that they weave with threads of (ecological) nature on one hand and (social) Nature on the other.

WHY MYTHOLOGY?

The importance of mythology here is twofold. First, myths highlight the weaknesses of control as an ideal and as an all-encompassing theoretical concept. Second, myths are instrumental in imbuing salmon with meaning. My use of the term initially emerged from my conversations with biologists. Williams's reference quoted near the beginning of this chapter was the first time I heard a biologist use "myth." Others later mentioned myths in different contexts, and there were numerous discussions of phenomena that can best be described as mythmaking practices and events.

MYTHOLOGY AND CONTROL

Mythology is noteworthy in part because it points toward factors that delineate the limits of Weberian rationality. Mythology does not so much erode rationality, especially control, as it hints at rationality's failures and its incompleteness. In particular, myth often reflects *nonhuman* control: what is beyond human social direction and manipulation in any meaningful sense. Without doubt, humans have been remarkably successful at extending their control of Nature—witness the societies that have been established in the climatologically extreme regions of the planet. Yet we have managed to push back Nature—to control nonhuman control, in effect—only a fraction. We still cannot breathe water or live long without drinking it. Our technological systems remain open to unforeseen factors and even to the power of those that we can anticipate (wind shear and its effects on airplanes; earthquakes and buildings). And we continue to die despite the best efforts of modern medicine.

Where rationality ends there exists *uncertainty,* a topic we take up below as one of three key aspects of mythology. It is the most important of the three, and it underlies the others as well: "bad science" and observer-created reality. Uncertainty highlights nonhuman control. An example of the control-limiting workings of uncertainty emerges from quantification and calculability. Modern biological research depends upon quantification to achieve its goal of becoming the next engineering. Yet often there are too many variables, each with too many possible outcomes, for biologists to claim much more than the beginnings of quantified control over many aspects salmon behavior, much less over the ecosystems in which salmon exist.

Biologists still seek control; they must in order to conduct biological research. But chance, one component of uncertainty, pervades biologists' worlds. In his discussion of the myth of Nature, Neil Evernden wrote that the "*overcoming* of chance thus becomes our central obsession"[7] in rationalized, industrialized societies. This point is of crucial importance here. The overcoming of chance *is* control. Chance implies that something exists beyond the limits of control. Biologists' R^2s, sometimes so low that even sociologists would find them disappointing, evince the overwhelming presence of chance effects. Control, whether it be over the salmon or over a scientist's own research agenda, becomes shadow boxing—or shadow grasping. Questions mount exponentially while answers do not even accrue arithmetically. Scope statements multiply, and with them universal facts become highly framed and delimited.[8] Knowledge as the predominant theme guiding inquiry gives way to uncertainty; the certainty of quantification yields to the uncertainty of qualification. Uncertainties also become myths when powerful interests direct that unknowns be remedied, that "gaps in the data" be filled when no data exist. "Expert systems" are created to fill those gaps, and "best guesses," highly educated intuitions but nevertheless intuitions, become the stuff of scientific fact and of public policy.

In rare instances the myths that are a part of salmon biology have been deliberately created; for example, advocates of the Canadian enhancement program argued that hatcheries were "proven technologies" even though they lacked any basis for saying so. The myth of hatcheries as proven technologies was foisted upon skeptical, even incredulous, salmon biologists. Regardless, it had important impacts on their work, owing to the social forces that supported the hatchery concept.

Most often, though, myths reflect biology's story, its understanding of the world. Biologists argue that those understandings are "facts." Skeptics point out that those facts are, given time, likely to be malleable and impermanent. After all, salmon biology's story of the salmon has changed over the years, and it will continue to evolve in the future. It may grow in sophistication as further rationalization of the salmon through high-tech methods improves and sharpens biologists' theories. But the point is that the new knowledge will produce new myths. Uncertainty is never far from the forefront.

MYTHOLOGY AND MEANING

The second reason for using the mythology concept here is that myths project meaning onto salmon. Whether it be "bad," fundamentally flawed science or "good," acceptable science (as distinguished by scientific convention), science creates its own myths and contributes to myths among the lay public. For instance, I remember a television documentary about salmon from my childhood, my first memory of salmon. I was amazed that the fish could find their way "home," back to the streams where they were born. *How* they found their natal streams was the stuff of social and scientific myth. Biologists thought smell or perhaps the position of stars was the cue. Recently it was found that salmon can detect differences in the earth's magnetic field, and biologists now feel that this plays a substantial role in the salmons' homing abilities. No longer are the salmon possessed of extrasensory talents or supersensitive noses; now they are tools, compasses with fins. Through science, salmon are mythologized, not in the sense of being founded upon false beliefs but in the sense of being constructed as part of a belief system, an interpretation of reality. Salmon have been constructed as scientific objects and as knowable and certain. Their meaning is like that of any Natural entity whose innermost, and most awe-inspiring, mysteries can be laid bare by science—a science that is itself mythical.

The next section of this chapter explores contemporary understandings of the concept of myth. Following that, I discuss the role of myths in social control. I then show that mythology in salmon biology is manifested in three forms: uncertainty, bad science, and observer-created reality.

CONTEMPORARY INTERPRETATIONS OF THE MYTH CONCEPT

In his survey of the constructions of Nature across disciplines, geographer Ian Simmons wrote, "Myths about the environment may be one of the inputs to our construction of it."[9] Others say that "myth" is virtually synonymous with "meaning." Foremost among them is Roland Barthes. Myth is not "superstitious or erroneous belief, or . . . primitive cosmology, but myth [is] an accepted story of the way the world is," according to one interpretation of Barthes' work.[10] A semiologist who wrote extensively about mythology, Barthes wrote in terms particularly apropos to considering myth and Natural constructs, noting that while

> there are formal limits to myth, there are no "substantial" ones. Everything, then, can be a myth? Yes, I believe this, for the universe is infinitely fertile in suggestions. Every object in the world can pass from a closed, silent existence to an oral state, open to appropriation by society, for there is no law, whether natural or not, which forbids talking about things. A tree is a tree. Yes, of course. But a tree as expressed by Minou Drouet is no longer quite a tree, it is a tree which is decorated, adapted to a certain type of consumption, laden with literary self-indulgence, revolt, images, in short with a type of social *usage* which is added to pure matter. . . . [M]yth is a type of speech chosen by history: it cannot possibly evolve from the "nature" of things.[11]

Barthes's use of the concepts "open" and "closed" here is telling, not least because of his acknowledgment that opening—giving voice to—that which was previously silent and without myth or meaning is a *social* phenomenon. This resonates with the discussion in Chapter Four about open and closed technological and social systems. Meanings become malleable and uncertain the instant that they become meanings in the first place, for sociological meaning is an open, shared concept. At the least it is slippery and often is the stuff of social conflict. Yet social groups also seek closure on meanings—they seek control over them. (It is also important to note that Barthes did not intend for his use of terms connoting verbal means of relating meaning and myth—"talking," "literary," "speech"—to imply that myths were exclusively verbal. He mentions that other means of communication such as photography, drawings, ritual, posters, and diagrams serve the same purposes; the *image*—in sociological terms, *symbol*—in myth is his central concern, whether the construction is cognitive or physical.)

Neil Evernden rightly notes that "the danger of myth is that it will be taken not as a human creation but as an independent entity existing outside the realm of culture. It will be perceived, in other words, as nature, as a 'factual system' when it is actually a 'semiological system.' "[12] Similarly, Ian Simmons wrote, "Myths can be one class of interactions between human societies and their non-human surroundings. [Niklas] Luhmann points out that ecological systems cannot communicate directly with humans and so our cognition of these systems and their behavior is a result both of such direct contact as we may have and of all the other channels of communications about environmental matters."[13] What bears special attention here is the role of social forces in the creation of myths. Evernden notes the importance of social interaction and social structure in myths of Nature, commenting that "what is significant for our purposes is [Barthes's] contention that, in this process, social concepts are secreted in nature, resulting in a new or pseudo-nature that is, in fact, fully historical."[14] The quest in this chapter is to explore the processes that result in this socially created, rationalized salmon myth.

MYTHOLOGY'S CONTRADICTIONS

Myths have a dual and contradictory character, an accidental quality to them. Often they are beliefs founded upon uncertainties, mistakes, and failures, areas and instances where control is weak. But myths may also indicate where control is at its strongest. The taken-for-granteds of a given social group may be seen as constituting a myth. Such myths, core beliefs, are enforced by a variety of implicit as well as explicit control mechanisms. These myths further control behavior because they often are perceived as being so powerful as to be beyond challenge. Perhaps the most common, and certainly clearest, examples of such myths come to us from the study of religion.

UNCERTAINTY AND MYTHOLOGY

Owing to a number of social and nonsocial "systemic" factors, uncertainty pervades many of the biological sciences and many complex systems generally.[15] Salmon biology is no exception. Yet it is a common perception that science *is* certain. Klaus Huberman, an especially widely read and philosophical biologist, told me, "One of the difficulties that you have in sociology is similar to what biologists find with ecosystems: they're so complex and so interrelated that we wonder how we can talk about it. Physicists had that problem, too, when they were dealing with phenomena within the atom. They finally found out that normal language wasn't sufficient." Physicists created their own vocabulary to communicate about the bizarre subatomic world.

Biologists, however, did not seem to me to have the physicist's understanding of the complexity of the phenomena that they try to understand, any more than sociologists have a comparable understanding. They face too many uncertainties to grasp the extent of the complexity of systems of interest to them. For example, previously I commented on the confident, upward-sloping lines of a Corps of Engineers graph depicting the return of salmon to the Snake River. The computer model created by the Corps generated numbers that resulted in scenarios of more and more fish returning to the spawning grounds. Only when I found that the unexplained variance was as high as 60 percent did I understand how uncertain those models were. The uncertainty was a product of complexity.

Uncertainty bounds the control that salmon biologists are able to wield over the salmon. The impact of uncertainty upon the other aspects of control identified in Chapter Four—control over salmon biology by structural forces and biologists' attempts to control other humans—is less clear. However, at this point a tentative hypothesis of sorts suggests itself: Effective control is inversely related to the level of uncertainty as expressed by those attempting the control. When science must be undertaken in surroundings of high uncertainty, myths are likely to be one product of the scientific endeavor.

UNCERTAINTY, EXPERTISE, AND MYTH

In most of my interviews I asked scientists what was the most important thing about salmon that was not known; in essence I was asking them where the greatest uncertainty lay. Most biologists answered by referring to the salt-water phase of the salmons' life cycle, where the fish spend 50 to 80 percent of their lives. George Williams put it this way:

> We really don't know an awful lot about the biology of salmon at sea. We really don't. The predictions about the abundance of salmon—our predictions are going to be 31 million pink salmon in

> the Fraser River this year or whatever—they're pretty Mickey Mouse. They're based on what we know about the numbers that went to sea. Sometimes we don't even know them, we just know the number of adults that spawned. You've got the historical records, so you can do a little bit of a stock recruitment diagram. But it's wide, wide variability, which is associated with variations in marine survival. We don't have a very good handle on that. There's a lot of speculation, but it's mostly beer talk.

Williams's response points to the biological and economic importance of remedying this uncertainty, according to the fisheries paradigm. For instance, unless salmon runs are made more predictable, there is the possibility of too many fish being caught, thus temporarily increasing economic return (assuming relatively stable prices) but in the long run destroying the runs and economic fortunes. On the other hand, poor predicting may mean that too few fish are caught, possibly flooding hatcheries and spawning grounds with excess fish and denying profits to fish processors.

Although salmon behavior in the oceans was the most frequently mentioned unknown, probably because of the simultaneous scientific and economic interest in reducing those uncertainties, other biologists mentioned issues unique to their interests that often were of little obvious value except to science. Neil Self, for example, said, "I think there are some things that we haven't got very far with. . . . We have not got ways to look at what salmon do at night. We have very poor understanding of what fish do in the dark, and we have poor understanding of what they do in the winter, and we have the worst understanding of what they do in the dark in the winter." Other biologists mentioned "carrying capacity," "habitat," or whether enhancement programs were doing what they were designed to do (increase productivity), remarks reflecting a variety of biological and economic perspectives on uncertainty. These diverse answers, and the depth to which some biologists went in explaining the importance of the uncertainties in their field, indicates the age-old paradox that *uncertainty is possible only with knowledge.* Knowledge, supposedly the generator of certainty, in practice also yields the opposite. It prompts more questions.

Adding to the uncertainty is the possibility that the knowledge in hand may not be as much about a particular species or phenomenon as about the broader social and historical processes in which a researcher works. One biologist commented, "One of the buzzwords from the '80s was, 'Expect the unexpected.' No matter how well you predict forward, something will come up and change everything. You can predict forward a wonderful model of rebuilding Fraser sockeye or whatever, all the model's parameters are right, and then some major natural thing happens that totally changes everything." The certainty of knowledge is tempered by the uncertainty that is generated as a part of the knowledge-gathering process. Those who accept the contingent character of the future recognize that their predictions—the illusion of certainty—are themselves attempts at mythmaking.

In its ideal form the scientific method implies that prediction—the ultimate, complete reduction of uncertainty—is supported by and even achieved by replication, repeating a prior experiment or study under the same or similar circumstances as a check on the validity of the theory guiding the study. When replication is not possible, scientists find themselves filling gaps in existing knowledge with what they acknowledge is just intuition. At least one pair of authors has labeled approaches that do not allow for replication "nonscientific."[16] Doubtless, many biologists would object to their studies being so labeled. Yet this gap-filling is a growth industry brought about by the very uncertainty that it is intended to remedy; it is an invention created by the necessities of rationalization.

Responses to these social needs—actually, efforts at control—are the primary way that myth creation emerges from uncertainty. Gary Hite's experience as a modeler bears out this assertion:

> A lot of the problem with modeling in the environmental/resource management context is, it's virtually impossible to do a repeat experiment. It is sometimes possible, and it is worth trying to do properly—management experiments where you have replicates and controls and the whole bit. But normally that's very difficult and people don't do it. The other thing is, a lot of the data is very noisy, very noisy. . . .
>
> It's not that you're very sure about the exact slope; it's that you're not sure about the fundamental things, like what is the carrying capacity? The data doesn't tell you because we have no data out here where we want to go. So in a lot of these resource management and fisheries things, the model is not so much an analytical tool that's

> based on hard data; it really represents a collection of the best understanding and ideas of the scientists who put the information together. That's an expert system when you think about it. . . . When there's not much data you think the lines will go up straight and flatten off. You know that's wrong, but it's all we can assume. We don't have the data to do any better.

The uncertainty—represented by "noise," or confusion in how to interpret the data—is not dispensed with. Instead, it is captured and quantified in the form of numerical assumptions. The creators of mathematical models deliberately incorporate scientists' "wrong" best guesses.

Harry M. Collins has written that expert systems like those created by Hite extract knowledge from so-called experts in a field and store it in a computer. "There is little of the deductive, model-building approach to this," Collins writes; "it is essentially pragmatic. The expert systems designer modestly accepts that we do not know much about what we know. . . ."[17] In a sense, expert systems encapsulate myth: uncertainty and the biologists' intuitions regarding how to dispel that uncertainty become intertwined. The expert system is a story about reality as salmon biologists *believe* it to exist. As Collins writes, designers of expert systems attempt "to render culture visible,"[18] exposing the experts' beliefs. Expert systems may make myths explicit, and in so doing those myths may be made available for debunking and revision. Such a self-reflexive stance is being taken by increasing numbers of biologists in contexts other than modeling as well, as the section on "observer-created reality," below, indicates.

Myths and "Bad Science"

Among salmon biologists the most obvious myths and the most troubling ones are those that emerge from "bad science." Bad science results from any number of faults, such as methodological flaws, poor analysis, lack of objectivity, or the absence of peer review. All of these were implied in George Williams's comments near the beginning of this chapter referring to "gray literature." Bad science, in his terms, is C-minus work.

Good science is ultimately the responsibility of the researcher, according to salmon biologists. None of the faults of bad science rest with Nature or with any nonscientific social phenomenon. Even the nonsalmon biologists brought into an agency that could have hired scientists specially trained in fisheries biology, as is alleged to have occurred in Canada, are ultimately responsible for the quality of their work product, according to the scientific ideal implied by Williams.

That said, no definitive definition of bad science emerged from my interviews; I take bad science to be defined as the opposite of "good science," and there was no consistent definition of good science among my research participants, either. When scientists spoke about bad science, in virtually all cases they were not referring to fabrications of data or any other form of scientific fraud or deliberate misbehavior. Although the first example of bad science below does have undertones of misbehavior by scientists, in general bad science refers to *poor* or *outdated* science, that which does not coincide with scientists' expectations or with their current constructions of a phenomenon: their myths.

FUNDING AND BAD SCIENCE

In the earlier discussion of the professional politics of funding salmon biology, I noted that, in the midst of a politically charged climate created by large sums of money being made available to large numbers of researchers, effective networks—rather than provocative ideas—may be the crucial factor in obtaining funding. This rebels against a core tenet of modern science, the "objective" appraisal of scientific work (including proposed work) by a scientist's peers. Such assessments are essential to modern science's mythology and its mystique. One biologist noted a phenomenon even more disturbing to those who embrace the scientific ideal:

> Even further, people disseminate proposals. If I were to submit a proposal for funding, that wouldn't be considered proprietary. The ideas in that proposal could be passed to somebody else who could then take my proposal, rewrite my ideas, and get it funded.
>
> . . . . It's very common practice in this kind of contracting situation. It's highly unethical, and within groups like the National Science Foundation it's like a career-ending "no-no" to do that sort of thing.

> In fact, even if you review a proposal for NSF or NIH, just by reviewing it you have to sign a document that states that you will destroy that proposal, you will not use those ideas. It's big-time stuff. You submit a proposal to Corps of Engineers, BPA, they'll xerox it and send a hundred copies out to the fishery community.

Logically, such behavior should come under the rubric of bad science, yet it occurs in peer review—that is, in the process of conducting good science. Biologists expect that the bureaucratic entities within salmon biology or that interact with professional biologists will follow scientific norms, not violate them as the biologist above asserts they do.

DISTINGUISHING FACT FROM BAD SCIENCE

When represented as "fact" (knowledge produced according to the guidelines of good science), bad science (again, that which is of poor quality or that has been superseded by more recent knowledge) may serve a variety of purposes. For one, as with any fact, bad science tries to simplify complexity. Biological systems are daunting in themselves, and in the case of the salmon they are poorly understood. When large-scale control over a Natural system is attempted, as in the case of the damming of rivers in the Pacific Northwest, technological and organizational systems interact with Natural ones. The result is a leap in complexity amidst what already are enormously complex, interacting systems, and there is a marked rise in uncertainty on the part of biologists.[19] This provides an opportunity for myths to be propagated. An example comes from Pam Mitchell, who commented on the migration of young salmon to the ocean in the context of the dammed Snake and Columbia Rivers:

> What we know is there's tremendous flexibility for fish; what we don't know is what their limits are and how far we push those limits. We don't know. Idaho Fish and Game, one of their biologists, he told his people that if those juveniles didn't make it from the place that they were born to the ocean in thirty days, that was it, they were dead. There's absolutely nothing anywhere that supports that. There are no data, no research that says that's the truth. In fact, National

> Marine Fisheries Service years ago had those fish a long time after they should have been at the ocean, and they would run them through a series of tests where they brought the salt content up and brought the salt content down and then back up again. The fish did just fine. They adjusted. Lab situation, artificial? Yeah. What does it mean? I don't know. But it does tell me that some of those people who say, "It's got to be this way," that we don't know that it has to be that way. I just want to do what's right for the fish, and I get so mad at the politicians and the ignorance.

Among salmon biologists, the implications of casually accepting bad science's facts as truth rather than recognizing the possibility that the information may be unreliable are staggering. In this case, the impacts include billions of dollars being raised through increased utility bills to spend on improving fish bypass facilities or other means of increasing fish survival. Similar amounts could be lost to the Pacific Northwest's economy if dam drawdowns end barge traffic on the Snake and the Columbia for months at a time.[20]

Mitchell's quote implies that managers such as those in fish and game departments are hard-pressed to create new, reliable myths. As the decision makers in matters regarding fisheries, they look to scientists to reduce complexity, not realizing that scientists often do not have the answers they seek. For example, one of the most important aspects of applied salmon biology is "stock assessment," the calculation of the number of adult fish available for fishing (the remainder being the "escapement," fish that return to their home spawning grounds or hatcheries to spawn the next generation). Managers rely on accurate stock assessments. Ian Smyth told me that, despite decades of stock assessment studies, "what I see is that when I sit down and crunch a lot of information to do a stock assessment, I can honestly say that I have never done a single stock assessment where all the information has been there and has been useful. What we always find is that when we sit down to use all the information and make up a coherent assessment or picture, you find that pieces of information have never been collected. You have to redesign programs." Complete information, "all the facts," may not be possible to gather. This is not because scientists are incompetent or because they do not have the necessary resources at their disposal. Salmon biologists are exceptionally talented, creative, and insightful researchers. And assuming that the funding is available—and I was led to believe that it is for efforts as economically important as stock assessment—assembling information for recognized variables is an accomplishable task. But the work of Smyth and others falls short of the ideal because of the constant presence of uncontrollable, uncertainty-producing forces.

Managers also create and propagate myths based upon bad science when, by attempting to exert further control over a salmon system (be it in a hatchery, a river, or the ocean), they only add to the complexity. Marc Raymond, who works for a large corporation with a tangential yet strong interest in salmon, told me,

> From my perspective, good biology is synonymous with good science. It's just a subcategory of it. And good science is dealing honestly and in a forthright way with the data, making decisions, forming your opinions based on what the information tells you, regardless of whether or not it happens to fit very well with your own personal set of beliefs. I feel very strongly that our area of research is filled with half truths and opinions about management that are not based on good science. But they have found their way into the management dogma. Once they're there, they are very hard to dislodge. I think we are as much fighting some of these old beliefs right now as we are generating new information and trying to understand how the salmon are being affected. So good biology to me is simply rigorous application of technical knowledge, of well-founded technical knowledge. And I quite frankly think to some degree we haven't been very good at it. . . .
>
> It happens in science as well as management, but I think it's a lot worse in management.

Raymond's job brings him into frequent contact with public-policy decision makers. His comments demonstrate both his skepticism regarding their ability to use information completely and wisely, and his perception of science's shortcomings.

Managers find themselves making decisions based upon information that is incomplete, of low quality, and that is contested by social groups—including by biologists. One biologist referred to managers as "people who are using their best judgment, but they don't have the background to really understand what's going on and what might need to be done." Moreover, as another biologist implied, bad science correlates with bad management; combined with lax managerial practices, bad science results in untenable myths: "By not sitting down and looking at it in terms of what are your management objectives, you go along on old assumptions based on flawed analyses if you are missing this information." The control-oriented objectives central to management practices are rendered impossible in a world of multiple, interacting social, technological, scientific, and ecological systems, what Charles Perrow termed "interactive complexity."[21]

BAD SCIENCE: SOME EXAMPLES

It would be misleading in the extreme to assert that salmon biology is rife with bad science. Still, as indicated by several of the preceding statements, bad science is pervasive enough to be a concern to scientists. Several more examples demonstrate the mechanisms by which bad science becomes transmitted from scientist to scientist, highlighting the sociological foundations of this phenomenon.

The first example does not refer to a specific incident but to a type of behavior. Pam Mitchell told me, "Not very many researchers like to write all the details down, especially if any of the details may make their study more questionable. They've been paid good money to do this. They don't want to look bad. And some of them simply don't consider the details important at the time. So it doesn't get written down, and when somebody picks it up later, they look at a number and say, 'Okay, this number must be right.' None of the qualifying factors are down with that number." Replication is a hallmark of the classical science paradigm, making detailed research procedures imperative. Yet Mitchell argued that researchers' presentation of self may interfere with the creation of replicable experimental protocols. For face-saving reasons researchers do not want it known that their findings hold only under a narrow set of conditions—that they are limited by severe scope statements.

Mitchell also alludes to the structural factors that may play a role in this phenomenon. Her mention of the importance of funding indicates that the rationalizing economic and political forces in salmon biology may compel such bad scientific behavior. If a researcher has cut corners, or if research that once appeared promising proves to be less than fruitful as time goes on, funding for related future research—or potentially for unrelated research conducted by the same researcher—may not materialize. Thus, there are structural, controlling pressures to present research results in the best possible light.

The second example of bad science comes from industry biologist Marc Raymond. In the same portion of our discussion in which he argued that management is replete with "half truths," he offered the following comments:

> Another good example, and this one is more topical, is: What is it that has impacted coho salmon? Clearly there are a lot of things that have been impacting them, but if you talk to a lot of the folks who are involved in coho research, the people who know it best, and ask them, "What is the single thing that they would do to recover coho?" . . . I think almost uniformly you would hear, "Reduce harvest. There isn't anything we can do until we reduce harvest to recover these fish." Yet we almost always, when we talk about these things, even at scientific conferences, focus on habitat issues rather than highlighting some of the harvest issues. I think some of these things get into dogma. They also become scientifically popular. They become sort of "in." And once they become hot, people run with them both because, obviously, people are interested in listening to it, they're getting a lot of positive feedback, and once these particular issues become institutionalized, if you will, they become popular subjects for funding. So there's a self-reinforcing process.

Raymond's comments regarding the institutionalization of bad science—"these things get[ting] into dogma" and becoming "in"—resonate with phenomenologists' idea of taken-for-granteds.[22] New ideas may find support among those who control resources, and they quickly become acceptable, sanctioned, and unquestioned.

The final examples of bad science leading to the creation of myths came from Will Perry, a biologist working for a fisheries-oriented state agency. For Perry, bad science and mythmaking have their roots in lazy scholarship. Bad science is the fault of all of those in a scientific community, not merely those who conduct the research of interest, in contrast to Williams's argument earlier in this chapter. An extended exchange between us took place:

> The other thing you've got to do with that information is you have to run down the original articles. Nowadays, the short cut is to review abstracts. But what you'll find in these articles is a lot of errors that somehow got through the editing process. So you can go out and do the quick and dirty thing, and the information that you use may not be correct. . . .
>
> *Q. Lore becomes fact, huh?*
>
> PERRY. Yeah. It's because some guy picks it up and didn't check.
>
> *Q. Is there one really prime example that comes to mind?*
>
> PERRY. . . . There's one study that was done at the University of Washington about twenty years ago. The guy's primary conclusion is that coho don't displace trout. It's been cited by hundreds of people. It's one of the primary references. What happened was I had other references that conflicted with it. So I looked up the guy's actual paper and looked at his actual numbers. His numbers showed that when you had trout and you put coho in—these were experimental channels—a couple of hundred trout left the area. They also took up a submerged position. Then he also cited studies that showed that steelhead in the presence of coho are mainly bottom feeders, which is completely different from all other trout. His own numbers completely conflicted with what he found in his conclusion. If you had looked at his abstract the way other people have done, you'd say so and so. But the actual tabulated results, which was just an appendix, was very different. That's real common. . . .
>
> There's another example, in a rather famous paper about territorial use. But the same author in a subsequent paper said he was wrong, that what he used was a laboratory setup that was too small and he in fact created an artifact. When he went back and did exactly the same thing with a much larger artificial channel, he got a different result. Some people never picked up the second paper. They got the first one. You've got to go back and cover the stuff that's out there. If you do that, you can succeed.

Perry's comments point toward two different processes of mythmaking united by a common theme. In the first, a researcher's poor data analysis led to incorrect information being disseminated throughout the profession. In the second, a researcher's poor experimental design led to the same outcome. That outcome is common, but the underlying factor that troubled Perry was the lack of serious initiative by biologists to consider in detail not only these papers individually, but the related research on these topics. In the first example, Perry's literature review revealed what he thought must be an inconsistency, an intuition ultimately substantiated by the author's own data. In the second example, the related research was the author's own retraction or correction.

In the academic and research milieu, status is the lone currency, and the surest way to increase one's status is to publish. Scholars and investigators are embedded in complex social systems. Varied, often intense, institutional pressures are a constant fact of their professional lives. When institutional pressures such as tenure and promotion reviews are added to the concerns over the presentation of research and of self exemplified by the first example in this section, researchers will make errors. When those errors are published, they simultaneously become bad science and the basis for knowledge in the field, the worst kind of bad science there is.

Observer-Created Reality

I conclude this chapter with another allusion to a problem first noted in the natural sciences by physicists, but one that is familiar to social scientists as well: "observer-created reality." Book chapters and even entire volumes are written about how researchers can minimize their interactions with the human objects of research.[23] However, it was not until physicists who were pushing the limits of the Newtonian paradigm reported that *how* they measured a phenomenon affected *what* they measured that the natural sciences began noticing this effect. *Scientists' tools, their methods and technologies, changed reality.* Physicist Heinz Pagels wrote that, within the post-Newtonian physics of quantum mechanics, "what an observer decides to measure influences the measurement. What is actually going on in the quantum world depends on how we decide to observe it. The world just isn't 'there' independent of our observing it; what is 'there' depends in part on what we choose to see—reality is partially created by the observer."[24] Put more forcefully, "Reality springs into being only when we observe it."[25]

A conscious awareness and acceptance of observer-created reality was not part of the reality of most of the biologists with whom I spoke. Klaus Huberman was an exception, however. He was an avid consumer of the lay literature conveying the latest developments in physical paradigms, from quantum mechanics to chaos theory and complexity theory. While he was fascinated with the potential applicability of these approaches to his field of study, he did not appear to have considered the appropriateness of the observer-created reality term until I asked him about it. Immediately he smiled and told me about the old days tagging fish in the Juan de Fuca Strait, then about his more recent work:

> They found that the fish, instead of going into the Fraser, went out again! It's called "re-forced migration" or "retrograde migration," and they attributed this to the fact that the animals were disturbed at a very dangerous time. We see the same thing in some of our sonic tracking; you've tagged the fish, now you've got to follow it. I gained my experience by tracking with someone with the National Marine Fisheries Service. I went to Puget Sound, met him at Whidbey Island where he had his boat, caught the sockeye salmon as they went migrating through Puget Sound, marked them with a sonic tag, and then followed them. Right. So we were off Whidbey Island, we catch a salmon, went through the whole procedure, tagged him, then released him, and for the next thirty-six hours we tracked this animal all the way out and lost him at Port Angeles! He never stopped. He just went out and out and out to sea. So, yes, there are lots of indications. If we can in sonic tracking, we like to ignore our first day's data. Every time you manipulate the animal, you affect it.

Observer-created realities add to the uncertainties that form part of the basis of mythologizing salmon, a part that is at least as important as are the certainties. Like Will Perry's example of the experiment in which the tank was too small to adequately mimic the stream conditions, biologists' tools may affect the fish and therefore their observations. One biologist put it plainly: "I'd say some of the stuff that we have monitored over the years has been more an artifact of how we did it than it was of the fish. So, yeah, the technology can affect the fish."

The lessons from quantum physics, and from the social sciences as well, indicate that researchers must be skeptical regarding the faith that they put into their observations. Weber was only partially right when he asserted that scientists "need no longer have recourse to magical means in order to master or implore the spirits, as did the savage, for whom such mysterious powers existed. Technical means and calculations perform the service."[26] Even with refined, quantitative, methods, observations still have a degree of uncertainty.

Some biologists spoke openly of this. In discussing the difficulties of studying fish in the winter, one of Klaus Huberman's Canadian colleagues said, "But as soon as you go in the water with the fish, you're influencing them. It's a real problem. So what you have to accept is the idea that you're probably going to learn a hell of a lot more about them by going in there and looking at them and risking some influence and trying to minimize it than you will be if you stand on top of the ice without a hole and complain because you can't avoid disturbing the fish." Researchers, like all constructivists, best proceed by being skeptical of themselves and their methods. Again, Huberman's colleague: "Clearly we interact and affect the experiments just by the fact that we think about things in a certain way. I've thought about attacking this problem in the lab in a certain way. The very nature of my approach to it may determine some of the things that happen, and maybe even bias the results, whereas someone else experimenting and approaching the same questions and the same experiments somewhat differently might have in that simple sense imposed a different set of conditions on the animals just by the way they've thought about it initially and perhaps get different results, perhaps have different effects on the fish." This biologist surprised me with his self-questioning. Reflexivity—self-awareness of the researcher's place in the research process and her or his impact upon that process—is not a common trait among biologists. Reflexivity places the mythmaker in the scene as part of the myth. For many biologists that violates their belief in an "objective" point of view wherein the researcher merely translates data from its raw form to a type that is accessible to others in the field. Salmon biology is undergoing a revolution of sorts, however, and some of the basic taken-for-granteds about the field are under assault, as the next chapter demonstrates.

CONCLUSION: INFINITE CONTROL?

Mythology emerges from our seemingly straightforward efforts to make sense of the world. It is at base a story about that world. Although in science's case the story is driven by rules more strict and detailed than those of grammar and punctuation, that pure, truth-seeking and truth-giving facade has cracks in it. Through bad science and observer-created realities, biologists create salmon that later scientists deconstruct. Uncertainty contributes to mythology by opening the way for "expert systems," sophisticated models programmed with educated guesses, to be included in biological, economic, and policy debates. Uncertainty also somewhat clarifies the limits of control that biologists seek to exert over the salmon. Control is not all that there is, even at an abstract level. What biologists' own myths highlight is the uncontrollability of so many of the phenomena that interest biologists.

In the prior three chapters control was the central concept of interest, whether the control was exerted over salmon, over salmon biologists, or over other social groups. This multifaceted control is the crucial reality—the primary social construction—of the human-salmon relationship. But mythology demonstrates some of the limits of the control concept. Control's academic champion, Jack P. Gibbs, argues that the potential for control is infinite, constantly expanding like Einstein's space-time continuum. Not only does such an expansive view of *any* concept challenge social science's skepticism toward the ubiquitous, but in this case it is fundamentally flawed. Gibbs's assertion is a holdover from an age when human potential, including the potential to control, appeared limitless. This perspective has been called "anthropocentric," human-centered,[27] and "exemptionalist."[28] As a species, our potential for control is not limitless. Physical, practical factors make this so, as does the presence of countercontrol phenomena. Mythology directs our attention to some of these anticontrol "forces."

Perhaps control is best considered an ideal type, an extreme and unrealizable example that is nevertheless instructive. An example of control as an ideal is the concept of a "closed system," discussed in Chapter Four. Schematically appearing as a closed loop, a self-contained linking of components without external, "environmental" inputs or internal losses (outputs), a closed system is realized control par excellence. Yet only cross-sectionally—in a snapshot of time—is any closed system capable of operating as designed. As anything other than an ideal, such systems are the equivalent of a perpetual motion machine; dreams of achieving permanent closure are just as hopeless. Like gravity or some other force slowing a perpetual motion machine, forces of Nature or of society impinge upon closed systems, cracking them open and revealing the fleeting quality of their alleged permanence.[29]

In admittedly open systems, such as salmon hatcheries, the control ideal slips further and faster because of the openness: water contains viruses from wild fish that spread rapidly in hatcheries; young fish "escape;" birds learn to fly in beneath protective netting over raceways and eat their fill before workers arrive for the day; adult fish "stray" to one hatchery from another; and complex technological systems fail, as when a worker cleaning a tank at a hatchery where "captive bred" endangered sockeye salmon were being reared accidentally left open a valve and literally flushed hundreds of juvenile fish down the drain.

Moreover, control does not explain mistakes, misunderstandings, misinterpretations, and uncertainties. Control may explain why it was that these things ever were able to occur, the circumstances surrounding mistakes and the like, but in the positivist's world of causality, it is not control that creates the failure. Instead, it is the problematical, uncontrollable factors integral to the milieu of salmon biology that do not simply limit control but militate against it. Uncertainty, bad science, and observer-created realities emerge as control creeps toward complexity. Myths congeal even as control melts away. Myths are the products of open systems. Anticontrol supplies the story's antagonists, enriches the plot, and makes the tale multidimensional.

Neil Evernden wrote that "Nature is *mythologized* from its inception: its contents are established through historical decree. There can be no exceptions: Nature is the realm of necessity, and there is no room for self-willed beings with purposes of their own. The conceptual purity of the domain of Nature is a condition for the security of the realm of Humanity."[30] In contrast to the dominant, anthropocentric view of the world, and even the universe, as available to human control, environmental activists, ethicists, and, increasingly, environmental *scientists* are questioning the extent of human control over our world: both its ethical correctness and the practical effects of the belief that infinite control is possible and practicable. Physicists argue that antimatter exists, the opposite of the primary particles that comprise all matter. Anti-control exists as well in phenomena like entropy, the second law of thermodynamics, that directs that all control is temporary. Myths embody entropy, as does freedom, the subject of Chapter Six.

6 Freedom and Self-Determination in Salmon Biology

WHEN PAUL MCGUIRE spoke at a professional conference about a survey of salmon biologists in the Pacific Northwest that he had conducted, I marveled at what he said. I had long been aware of the subdiscipline of biology known as "conservation biology," and a couple of biologists had come close to sounding some of its themes in our interviews. They said things like, "Salmon have uses beyond their economic value," and, "We've got to preserve them. Biodiversity tells us that." Yet no one so consistently or so thoroughly—so *adamantly*—spoke of the imperatives of conservation biology as they related to salmon.

Conservation biology, a young sub-field, challenges the dominant biological paradigm's directives regarding objectivity and engagement in ethical and social issues related to the profession. It stands in stark contrast to the perspectives guiding those areas of biology that have been most valuable to industrialism, such as forestry and fisheries. Conservation biologists argue for the "intrinsic worth" of all species. This leads them to assert that biologists have a "responsibility to the resource"—ethically and normatively, they *must* be advocates for their species of interest. This in turn compels some of them to assume the role of advocate and even activist; conservation biologists have filed Endangered Species Act (ESA) petitions requesting protection of butterflies and salmon, they have given testimony at numerous public meetings, and they have published articles in environmentalist magazines and newspapers that more traditional biologists would avoid.

McGuire was the first salmon biologist in my experience to clearly sound themes of conservation biology. His survey revealed that his peers felt strongly that great lengths should be taken to preserve the remaining salmon runs of the Pacific Northwest. They said that the weight of the research supported their contention that the salmon were biologically integral to the region and deserved a secure future for that reason alone. Months later when I spoke with McGuire in his office, he expounded on those themes. A map of his favorite research site was pinned to his wall, and framed certificates attesting to his abilities as a biologist and to his substantial contributions to the American Fisheries Society's regional chapter framed a large window looking out on a parking lot and dense woods beyond. After our interview, McGuire pointed with pride to a small vial of gravel and water. "I got that from Redfish Lake," he said. The Redfish Lake sockeye salmon run in Idaho was the first in the Pacific Northwest to be listed under the ESA, and it has become a point of tension among biologists. "Went up there last year. I had never been there before. I just had to go up and see it for myself. That's the gravel they spawn in." This seemed to be the salmon biology equivalent of a groupie getting a rock-and-roll singer's autograph.

McGuire's quiet admission that he was a fan of the salmon said much about his "identification" with the fish, to use his term for it. None of the biologists I had spoken with before, or whom I spoke with after, seemed to have so deep and absolute a commitment to the salmon, although others shared McGuire's vision. His statement of why he has taken a conservation biologist's stand for the salmon was the clearest and most forceful of the lot:

> I care about the animal. It's an intimate, integral part of the Northwest ecosystem. When you think of the Northwest, you think of timber, you think of salmon, you think of water. It's a keystone resource for the Northwest. We as a population really haven't given it the respect that it deserves. I've seen beautiful salmon runs, and I've seen their ebb and flow with the weather, with the climate, with the terrain, and how they fit into the system. It's a marvel to behold. I look far beyond the economic value, as you can tell. So over many years and a variety of studies, I've just gained such an awe and appreciation for their resilience and for their adaptability, their variability, and their beauty—their pure and simple beauty, visually and in regards to their life history. . . .
>
> I've given a lot of thought to them and how they all fit together, how they've evolved. You can't help but gain a great appreciation for them as a fellow organism that fits into the Northwest. I feel so sad-

> dened when I see the state of our rivers. The Columbia River is such a great example of what we've done to the tremendous stock of salmon that ran through that river, the genetic variability that existed there pre-1920s, 1930s.

At the time I interviewed him, McGuire's language was rare in my study and in ways was unique. I recall no other biologist before him describing salmon as "beautiful" or admitting that they stood in awe of the fish.

McGuire's was a construction of salmon that, from my experience, was clearly in the minority but that reflected in contemporary terms most of the history of human social interactions with salmon. Traditional Native American constructions of the salmon, for example, were and are similarly reverential. Historically, tribes in the region welcomed the "salmon people" back each year with ritual and feasting. After feeding the year's first salmon to the village's children—a sublime act of socialization—they set the fish's bones back into the river from whence it came so that it could return to the fish people in the depths of the ocean and affirm that the people of the land were properly respectful.[1] Salmon were "fellow organisms," mysterious, full of life, and givers of life.

This chapter argues that the conservation biology perspective is the most important of several emerging trends in salmon biology that hold the possibility of substantially revising the discipline's dominant, use- and control-oriented construction. The upshot of the perspective is decidedly normative: that both biologists and salmon should be allowed more freedom. The following section treats freedom as a polar opposite concept of control and, as was the case with mythology, I consider freedom to be a means of distinguishing the limits of control. With that as the basis, I present the conservation biology perspective as found in salmon biology.

FREEDOM AND CONTROL

FREEDOM IN CLASSICAL SOCIOLOGICAL THEORY

In the opening pages of his study of freedom, sociologist Orlando Patterson wrote, "There are remarkably few general histories of freedom, and those few amount merely to a record of the ideal in different periods without any serious attempt to explain why the commitment persisted throughout the millennia of Western history. . . . The most stunning characteristic of the history of freedom is its continuity. Freedom . . . emerged as a supreme value over the course of the sixth and fifth centuries b.c., at the very dawn of Western civilization."[2] Despite freedom's millennia-long presence in Greek, Roman, European, and U.S. history, within North American sociology freedom today is but an occasional theme, hardly the stuff of intense study.

The work of the earliest theorists in the field yields a much different picture, however. Jeffrey C. Alexander notes that freedom was an underlying concept in the work of the most dominant classical theorists. He wrote, "The central problem in Marx's work, all agree, is the relation between freedom and necessity."[3] In Marx's writings freedom is an attitude defined as "the *feeling* of man's [*sic*] dignity."[4] Freedom was not a behavior but a mental condition.

Emile Durkheim's approach to freedom was more complex than Marx's. Alexander wrote that his "commitment to national integration and social control was counterbalanced, in Durkheim's mind, by an equally strong commitment to the democratic expansion of liberties for individual citizens."[5] Freedom existed only by virtue of control: control of the state, which, Durkheim wrote, "is extending itself further and further without the possibility of assigning it, once and for all, a definite limit."[6] Freedom and control stood in opposition to one another, a tension that Durkheim argued was being increasingly relieved by the hegemony of society's controlling forces. Despite his reading of historical forces, Durkheim held firm to a concern for the individual and for individual freedom in the face of the state, whether that state be capitalist or socialist in orientation.

As is the case with Marx and Durkheim, Max Weber stands squarely in the enlightenment tradition of reason, with "its assumed capacity to promote human freedom."[7] Alexander wrote that "Weber's theoretical achievement is enhanced by . . . its driving ideological concern for freedom and individual control."[8] Thus, for Weber freedom and control were interrelated, not in opposition. "Free action depends upon a control that is achieved when an individual purpose is formed over and against the barriers of both affect and conditions. 'An actor's "decision" is "more free" than would otherwise be the case,' Weber writes, to the degree that it is 'based more extensively upon his own "deliberations," which are upset neither by "external" constraints nor by irresistible "affect."' "[9] Weber's instrumentally rational approach posits a dominant normative order as a prerequisite to freedom; society's values control and direct what is considered free by establishing standards that are not dependent on immediate circumstances but instead possess a more permanent quality.

Control is also distinct from and opposite to freedom in some of Weber's writings, however. For example, although not directly addressed by Weber, the norm of control over Nature through science appeared to have been a salient topic when he wrote, "The historical development of modern 'freedom' [in Europe] presupposed a unique and unrepeatable constellation of factors [including] . . . the conquest of life through science."[10] The "conquest"—control—"of life through science" was one of the factors that enabled "freedom" to emerge; the two have a distinct, conflictual, even dialectical, relationship.

Freedom was a topic of special concern to some of the earliest, and still most influential, sociological theorists. Although they were far from being of one mind regarding the concept, some commonalities are notable. One is the tension between freedom and the collective good, expressed as "control." Marx, Durkheim, and Weber each indicated that slavery or the near slavery of serfdom was crucial to the development of the Western concept of freedom. So, too, did Patterson, who argued that historically "freedom was generated from the experience of slavery."[11] The imposition of an extraordinary amount of control over persons' lives led to the construction of freedom as an opposing concept.[12] Additionally, Durkheim and Weber identified in capitalist industrialism a strong impetus toward social control as against individual freedom, while Marx stressed the control exerted by dominant social classes over the freedom sought by subordinate classes. Control and freedom as theoretical concepts, then, have a long and conflictual relationship. That conflict is displayed in salmon biology today, and I discuss it in detail below. First, though, it is necessary to explore the clash of these concepts in the abstract.

CONTROL/POWER VERSUS SELF-DETERMINATION AND FREEDOM

In the foregoing discussion I have not used control and freedom as rigorously as I have previously used control and its mental counterpart, power. Let us now reintroduce that rigor. As previously defined, control referred to behaviors and artifacts used by a social group to direct the behavior of other persons or nonhuman entities; power was the knowledge or belief that control is possible. In contrast, freedom and what I will call *self-determination* stand as conceptual opposites of power and control. Self-determination is exhibited by behaviors that are unconstrained by the actions of some other group, and freedom is the corresponding attitude by a social group that it acts of its own volition. Figure 1 locates these concepts on a spectrum.

As shown in figure 1, control/power and self-determination/freedom correspond to what might be called an other/self distinction. Put succinctly, from the perspective of social actor (A) control/power is possessed by another social actor (B). In contrast, self-determination/freedom is possessed by (A). In everyday life we do not distinguish so sharply between these terms, of course. We use concepts like power and freedom loosely, commonly referring to ourselves or a social group with which we are affiliated as having "power" and groups with which we are in conflict as being "free" to act. The object here is to identify and explain a set of analytically useful concepts grounded in the data that will enable us to gain a greater understanding of persons' social realities.

As an example of how I envision these concepts being used, consider an environmental group that claims that a certain salmon run is endangered due to overfishing. This would be interpreted as a construction of the salmon run's continued well-being, or self-determination, as threatened by the controlling behaviors of fishers. It is important to note that the same concepts may be used to interpret the other group's construction of the situation as well. That is, the fishers may argue that they will be unable to make a living if the environmental group succeeds. We would interpret their construction of the salmon as that of a vehicle used by environmentalists to control their livelihood and thus restrict their self-determination.

Locus	*Other-Directed* ◄——►	*Self-Directed*
Physical/Behavioral	Control	Self-determination
Cognitive	Power	Freedom

Figure 1. The Control/Power, Self-determination/Freedom Spectrum

Second, it is important to recognize that neither end of the spectrum portrayed in figure 1 can be completely achieved in human social interaction. Even in the most repressive regimes, some social groups manage to secure for themselves a degree of self-determination and freedom in society. Similarly, social norms and other aspects of social structure will always be present to inhibit the behaviors of those within even the most anarchistic of social groups.

I have introduced the control/power, self-determination/freedom distinction primarily because it emerged as an important dichotomy in my research. It also is useful because, while control seems so present in our lives, countercontrol—self-determination—is and has been crucial to our behaviors toward other social groups and toward Nature. Historically, control is a poor explanatory concept for human social relationships with the world in which societies are embedded. That is, in the distant past control simply was not an important aspect of many societies' relationships with some nonhuman entities. Certainly control was evident in the efforts of gatherer-hunters at obtaining vegetable or animal foodstuffs, for example, but those exertions of control were not attempted over large numbers of nonhuman organisms; that is, agriculture or livestock domestication was not practiced as a primary mode of subsistence. Indeed, the success of such societies depended upon the possibility of self-determination of nonhuman biological entities—their ability to forage, reproduce, and otherwise go about their business of living. Similar points could be made regarding nonliving entities (the importance of free-flowing streams, for example).[13]

In addition, as today's industrial mode of production sweeps to every corner of the planet, presenting humans with numerous opportunities to "play god," as one author has argued we have done in America's national parks,[14] biological scientists and laypersons alike argue that realizing society's potential for control over nonhuman entities may be a mistake. This construction of Nature asserts that less, rather than more, control may be called for, even to the point where control over nonhuman entities fades to self-determination for those entities. An example of this is habitat restoration programs for salmon that improve in-stream and streamside conditions previously degraded by humans, allowing salmon to reinhabit the area. In so doing humans are relinquishing control and extending the possibility of self-determination to the salmon, the stream, and all the related nonhuman components.

It is important to note, however, that even while humans relinquish control over, for example, a salmon stream, they are almost certain to have to impose new controls on human behavior in order for the nonhuman entities to achieve self-determination and, in humans' eyes, freedom. This observation only reinforces the logic of viewing control/power and self-determination/freedom on a continuum. Let us now explore the contours of that spectrum as it is found in salmon biology.

BIOLOGISTS' STRUGGLE FOR FREEDOM

THE SCIENTIFIC IDEAL IN AN AGE OF LIMITS

Salmon biology in North America emerged as a discipline out of governmental and economic pressures to rationalize the fish of the United States and Canada. Those rationalizing impulses have continued to guide the fisheries biology profession. Now, however, amidst fiscal crises for state, provincial, and federal governments, those rationalizing pressures are increasing. Demands for improved efficiency and productivity mount. Many biologists predicted that the result would be a reduction in the pace of knowledge creation about salmon. For instance, George Williams said,

> I think more and more of the research being done on salmon and on other things is aimed at answering questions that are of immediate importance to management, whereas thirty years ago or more you did research that had relevance for management, but what really motivated it was trying to find out something about the animals. The interest was in the salmon, not in the management per se. Here, this seesaw has gone on for years and years and years, with management saying, "God, you scientists aren't doing anything that's of help to us at all," the scientists saying, "Look, if you ask us to answer those types of questions, we're never going to learn anything."

Scientists struggle to secure the leeway to conduct studies as they see fit. They desire professional self-determination. In large part that means working on topics familiar to them. Yet productivity and efficiency in Canada's government laboratories create a trend antithetical to this ideal.

Williams angrily said, "I still believe to a large degree that people who work for government labs: Really there's no point in hiring somebody who is highly trained, well educated, and tell them what they ought to do. . . . In a lot of government bureaucracies people up above are always wanting to tell you not what's on their mind, but how they want you to solve it. *How.*" Neil Self, also a Canadian, sounded a similar note, commenting, "The scientists, I think, need freedom to do science. This doesn't mean by any means that scientists should just go do any damn thing that they feel like. What it means is that scientists should be brought into an organization, there should be some kind of dialogue with them to begin to see what interests they have, what talents they've got, and you begin to match them up with certain kinds of areas where you want to see things progress. You move them into it and then let them go."

These comments portray a normative ideal that persists among scientists. It holds that the proper scientific presentation of self is that of the independent, engaged, committed researcher. No external forces impinge upon the scientist's project, and the work that he or she undertakes is probably more a result of individual initiative than it is of instrumental need. But in today's political and economic climate, researcher self-determination shrinks as those institutions exert their control over research. Canada's situation is more dire than that in the United States, and it also has centralized more of its salmon research (and fisheries research generally) in federal government laboratories rather than disperse research to other locales as the Americans have done with universities and state governments.

In the United States the preponderance of salmon research takes place in university settings. Although funding is tight for all researchers, those in academia feel they have an advantage in research freedom. Alex Stand said, "That's probably why people like to work in universities, in contrast to government agencies or businesses. You may not have the guaranteed research funding that you may have in those environments. On the other hand, you've got more freedom to explore interesting directions. So I think that's a strong point in universities." Even Marc Raymond, who conducts salmon biology research for a timber company, felt he had a substantial amount of freedom:

> I would say, in general, that my research . . . has been generated as much by my own particular interests as it has been by being dic-

> tated by the company in terms of, "We need an answer to this particular question. Go out and get that answer." To their credit, I think they recognize that, because of the fact that I'm involved in research, I was probably more aware of the issues that would be arising and maybe could get a better handle on what we ought to be addressing. . . . Within limits, yeah, we have a substantial amount of freedom in terms of the research that we do.

The freedom ideal as played out in acts of researcher self-determination persists, although research depends upon funding, and the highly politicized funding process is a powerful manifestation of control.

That researchers believe they are free to conduct their studies as they will in the face of controlling factors of which they are fully aware points us to an especially provocative phenomenon, one that scholars of organizations have long since identified. Charles Perrow wrote that in bureaucratic settings "there is a never-ending struggle for values that are dear to participants—security, power, survival, discretion, and autonomy—and a host of rewards. Because organizations do not consist of people who share the same goals . . . and because control is far from complete, people will struggle for these kinds of values."[15] Resistance to the restraining impulse of control is, perhaps, not entirely futile.

More importantly, as Perrow implied, control and power are never complete. Within any control-oriented setting there also are impulses toward freedom and self-determination. In a bureaucracy an employee will "cut corners" to make her work easier or to assist clients. In a county jail, inmates will gamble because gambling brings a bit of pleasure to the inmates' lives. And salmon biologists will do things like bootleg initial research involving an unfunded project onto an existing, funded one. None of these practices is within the rules. Because the social norm of freedom and acts of self-determination are so pervasive in the West, power and control face constant challenges.

A demonstrable conflict exists in salmon biology between control/power and self-determination/freedom. Biologists are aware of this in their own ways, especially the Canadians working for the federal government. Klaus Huberman told me,

> We debate these things, obviously, and there are as many ideas as there are people. I think the best way probably is to make a deal

> with the scientists. Say, "We will allow you to spend 60 percent of your time on some major questions that you consider key questions that you identify, being a specialist in your area. But the other 40 percent, we will ask you to help us out with this firefight." We may do this in such a way that they could say, "We'll give you one or two years to do pure research, and then you'll spend one year trying some applied research." So there will always be 60 percent of the staff involved in, I wouldn't call it pure research, but understanding assumptions, really understanding. . . .

Many biologists say they seek such a compromise. They feel obligated to resolve questions of applied research and they seem uncomfortable making absolute claims that they should be free to explore questions of interest to them. Such claims sound unrealistic to them.

Those in Canada identify the financial difficulties faced by their nation as being at the root of their pragmatism. One biologist observed that the ratios of salaries to research funding at the Pacific Biological Station was probably 65:35. He added, "If the trends that we see now continue, it wouldn't surprise me ten years down the road to see that at 95:5 or at 85:15, which means that you will have seriously altered the balance of how investment is weighed to support research." The freedom to conduct research is severely limited when the expertise is there but the financial resources to support the research are not. Within salmon biology, both freedom/self-determination and power/control presuppose a need for *resources,* whether those resources are economic, technological, or otherwise. That need is a form of control. Thus, it can be argued that salmon biologists must control some resources to be self-determined.

THE IMPORTANCE OF INTERCHANGEABILITY

These comments lead to an important problematic, that of the limits of freedom. It was implied by George Williams, who, in speaking of his days administering large research organizations, said,

> I'm a pretty good salesman, and most good salesmen are pretty gullible, so I could be fairly easily persuaded to do what somebody wanted me to do. If they'd convinced me it was important, I'd do it.

> Other people who are very good researchers bristle. They want to do what they want to do. A lot of those people dig themselves into a very deep rut over the course of a career. They do their Ph.D. thesis on moles and then they do a post-doc with some other guy on moles and then they continue to work on moles. Working on gophers would be a giant step for them!

For Williams, too much freedom can be a bad thing. Although some biologists complained that they already were fettered, and many others saw a day when they would be unable to conduct any research that they felt was important, an anarchic approach to research was of special concern to Williams.[16] Williams believed that good scientists are flexible and interchangeable, able to easily shift focus from one subject, and even one species, to another. Such interchangeability allows researchers more freedom; if an opportunity presents itself to work on moles rather than on salmon (or if it becomes a necessity to do so), then a researcher should be able to make that shift. Researchers who strongly identify with a species or who otherwise resist interchangeability are, in Williams' eyes, something less than competent biologists.

This additional form of interchangeability deserves special attention. The concept was first introduced in Chapter Three, when I presented the concepts of the interchangeability of species and the interchangeability of salmon. Here, interchangeability is at the crossroads of control and self-determination. Williams told a story that exemplified his perspective. Early in his career he was offered a position in British Columbia: "I accepted the job, came back, got married, my wife and I came out here to the West Coast, and we went to the home of the head of the department. . . . I didn't know when we went to dinner whether I was going to be working on moose or trout. We got around to coffee and [the department head] said, 'I think you'll enjoy working on trout.' That's when I knew I was into fisheries."

Few of the other biologists with whom I spoke, and none younger than fifty-five or sixty, similarly saw themselves as so flexible that they could work on species so outwardly different as moose and salmon. Specialization in biology and zoology limits researchers to a greater extent than it once did. However, it appears that the needs of government, in particular, compel some decision makers to ignore this trend in the scientific community.

Specialization may be construed as self-determination, and the lack of appreciation of it angers some biologists. For them, interchangeability is anathema. Neil Self put it this way:

> It's much more of a contractual approach, contractual thinking. You're a scientist and I say, "We've got a problem with escapement of chum salmon or halibut population dynamics. You're going to work with that." You say, "Well, my background is all in biochemistry and biophysiology." And I say, "Well, that doesn't matter. You learn to do this. You're not just here to indulge yourself in the thing that you think is most interesting." That's the actual mentality of research administrators. And people are told, "You've got to be able to learn to change." But why on earth do you take someone like a scientist, who spent years getting his education and they build a very powerful platform from which he works, and then tell him to work somewhere else where he may not have the interest or the skills or the motivation?

If there is a sense of elitism to such statements, it emerges from the division of labor in industrial society. Scientists are told that generalists do not get jobs. Yet the pressures of economic rationalization—especially obtaining the most productivity from an agency's work force—overwhelm the scientific will toward freedom and self-determination. Control appears to be the dominant theme. However, as the following section demonstrates, self-determination and freedom for biologists and for the salmon themselves is an emergent, and potentially powerful, movement in salmon biology today.

Conservation Biology, Freedom, and Self-Determination

Fisheries biology stresses an anthropocentric, human-focused, approach to salmon. Anthropocentrism is characterized by a utilitarian outlook toward nonhuman entities: humans have a right to use Nature as they please. Variants of anthropocentrism such as "stewardship" and "wise use" assert that humans must use "natural resources" with care; otherwise those resources will not be available to future generations.[17]

Still others argue that stewardship is of minimal importance given the resourcefulness of humans at finding replacements for that which we have used up or driven to extinction.[18]

In contrast, conservation biology's perspective approaches *ecocentrism,* in which what is right and good is judged by its effects on ecosystems. In his essay "The Land Ethic," wildlife biologist Aldo Leopold penned conservation biology's guiding ecocentric edict when he wrote, "A thing is right when it tends to preserve the integrity, stability, and beauty of the biotic community. It is wrong when it tends otherwise."[19] Ecocentrism directs that humans are not the measure of all things. Rather, nonhuman entities should be vested with value entirely apart from any use that humans might make of them.

While continuing to embrace "good science" as a core tenet, many conservation biologists accept as another crucial component of their discipline the importance of protecting and preserving species, habitats, and ecosystems. For many of them this necessitates *advocacy* on behalf of those organisms and places. However, as we shall see, considerable conflict has emerged among salmon biologists regarding their proper role: as "objective" reporters of data or as "advocates" for salmon.

CONSERVATION BIOLOGY: THE CORE

The field of conservation biology has grown rapidly since its inception in the early 1980s. Several journals, including *Conservation Biology,* publish scholarly works, and conservation biology courses are taught at many colleges and universities. Paul R. Ehrlich, an esteemed biologist and one of the discipline's cofounders, wrote, "Conservation biology is the subdiscipline of ecology that attempts to develop sound strategies for coping with the crucial problem of the preservation and restoration of organic diversity."[20] Somewhat more urgently, Edward O. Wilson wrote, "Because extinction is forever, rare species are the focus of conservation biology. Specialists in this young scientific discipline conduct their studies with the same sense of immediacy as doctors in an emergency ward. They look for quick diagnoses and procedures that can prolong the life of species until more leisurely remedial work is possible."[21]

Ehrlich and Wilson's support of conservation biology should not be underestimated. They are two of ecology's most outspoken and widely respected ambassadors, men of immense status. Their comments above indicate several important points that are fundamentals of conservation biology. One is that, in terms of conducting "good biology," conservation biology sacrifices nothing. Indeed, its practitioners have been inventive, creating new research methods and developing theories to explain the phenomena of concern to them, including processes affecting biological diversity (the variety of plant and animal species in a given area) and the causes of species extinction. A second point common to both of these quotes is that conservation biologists emphasize not only species *preservation* but the *restoration* of extirpated, threatened, and endangered species as well. The subfield of "restoration ecology" has emerged as an important component of conservation biology.

Three other closely related aspects of conservation biology that play especially important roles in salmon biology can be added to this list. First, the establishment of conservation biology as a discipline appears to have been driven almost as much by an ethical perspective as by a biological one. Conservation biologists ask how it is possible for those scientists who know a place or a species best to stand idly by as extinction or destruction takes place. Ethically, scientists *must* get involved.

This leads to the second point: conservation biologists often translate this ethical imperative as one that compels advocacy of some sort. That advocacy may entail a range of undertakings, from lending support to environmental organizations to filing Endangered Species Act petitions.

Third, regardless of the mode of advocacy or activism, the aim is the same: to affect the public policy process. The conservation biology approach was provocatively summarized in an article published in the *Earth First! Journal*, the publication of the radical Earth First! movement, in 1987. Written by Reed Noss, later an editor of *Conservation Biology*, it read in part:

> Not long ago any scientist who spoke out in favor of protecting the earth was scorned by most of his/her peers. A few biologists who were already well-established in their profession, most notably Paul Ehrlich, were able to speak in behalf of environmental preservation as far back as the 1960s. . . .

> Respectability and credibility are of utmost importance in the bitterly competitive world of science. Any scientist who publicly displays affection for living things or any hint of an ethical commitment to saving them will experience difficulty procuring grants and academic appointments. Even within the science of ecology, which ecologist Paul Sears fondly called a "subversive subject," the ideal of pure, value-free objectivity demands that ecologists never admit to the social, political, or ethical implications of their subject matter.
>
> Fortunately for earth, this "professional" bias against involvement in conservation and ethics is beginning to erode. During the last decade, more and more ecologists and field biologists have grown concerned, for a very poignant reason: the study sites they know and love are being destroyed. Some of these scientists realize that whereas value-free objectivity is an appropriate though difficult ideal for the process of testing hypotheses through observation and experiment, the interpretation of scientific results, and even the positing of hypotheses in the first place, can and must involve other human capacities. Moreover, the direction that science takes with regard to the health of the biosphere can be either negative, for example, most industrial technologies, or positive, for example, studies of nature reserve design. . . .
>
> The increasing respectability of conservation biology in the scientific arena is a signal that biologists are getting worried. It is a healthy worry. Out of this worry has arisen the Society for Conservation Biology, in which scientists can mingle respectably with professionals in other disciplines to study the processes of extinction and environmental degradation, and devise strategies to avert these disasters.[22]

More recently Noss wrote, "I define conservation biology not only as multidisciplinary, but also as mission-oriented. I call it science in the service of conservation. . . . Science, as traditionally defined, does no good if isolated from 'softer' issues such as ethics, sociology, and political strategy. Indeed, there is nothing more dangerous than science in an ethical vacuum."[23]

Conservation biology constitutes a social movement within the broader discipline of biology, and within fisheries biology as well. Noss's phrase "science in the service of conservation" implies the overarching importance of this activistic, change-oriented quality, and conservation biology displays many of the characteristics of a social movement. It possesses resources and is organized, insurgent, and ideologically distinct from prior biological perspectives. Yet it maintains continuities with the past, including its emphasis on scientific rigor.

Within fisheries biology the movement reflects several of Noss's points. First, biologists openly identify with their research subjects and, second, they profess a moral responsibility to them, as Paul McGuire's comments at the beginning of this chapter demonstrate. Third, the salmon and the habitats that they require for survival are being destroyed at an incredible rate. For instance, one study published in the American Fisheries Society's journal *Fisheries* estimated that 214 "native naturally-spawning Pacific salmon and steelhead stocks in California, Oregon, Washington, and Idaho . . . appear to be facing a high or moderate risk of extinction, or are of special concern."[24] Finally, conservation biology is slowly, almost grudgingly, being accepted by fisheries scientists.

CONSERVATION BIOLOGY WITHIN SALMON BIOLOGY

Identification and Ethics

The American Fisheries Society's Oregon chapter, established in 1965, has long been a hotbed of controversy for its emphasis on "conservation." Based on Noss's writings and several of my interviews, it appears that the word *conservation* as conservation biologists use the term has powerful overtones of ecocentrism if not outright *preservation*—the setting aside of "natural resources" by prohibiting their use (in contrast to older definitions of conservation that imply use of resources). This definition of conservation emphasizes sustainable populations of species and stable habitat; it does not rule out some use of the species, land, water, or other entity in question, so in that sense it probably falls short of preservation. But it does direct that economic valuation of resources is a secondary valuation, and such an approach to nonhuman entities runs counter to the historical fisheries orientation to salmon and other species.

The effect of the Oregon chapter's open advocacy of conservation was to distinguish it from the mainstream fisheries perspective. As Bruce Alton, a long-time Oregon chapter member, told me, "I don't know if it's the distance from the Washington [D.C. headquarters] office of the Society or what, but it tended to be sort of a maverick or proactive, if you will, right from the beginning. . . . The Oregon chapter has been one of the leaders of environmental consciousness while being a voice for environmental protection and fish conservation." It should not be surprising, then, that the chapter has been a leader within the AFS in adopting a conservation biology perspective. Alton said,

> I remember a famous quote from Peter Larkin—to me it's famous—who's one of the heroes of West Coast salmon biology. It was about 1965—I mentioned this to him at the 1993 Portland [national AFS] meeting, where he received a prestigious award—and the last Portland meeting before was in 1965. He was a keynote speaker, and I think that may have been where he said, "Two generations of fishery biologists have tramped the streams of the Pacific Northwest documenting the decline of the Pacific salmon." I told him, "You can change that to three now, Peter." I think for too many years that's what we've been doing. We've documented the decline. But I suppose it's sort of Pollyanna-ish to say that early biologists should have done more. They were up against some pretty powerful horses and they had very little public support. Woody Guthrie was right: the people wanted jobs and they wanted power. They didn't want fish.

Alton and a growing number of others believe that the data are in and that it is time to do much more for salmon and for other fish. They are compelled to act because of their strong identification with the salmon. As Stephen Fogarty, one of the youngest of the participants in my study, told me, "The best biologists, the ones who are going to develop special insight, are going to be the ones to get intimately involved with their species or the broader ecosystems or whatever they're working with."

Ever since Descartes, science has been characterized by a subject/object divide, a belief that scientists have nothing in common with the objects of their studies. Many, no doubt the majority, of biologists firmly cling to this distinction. Yet as ecologists ponder the meaning of ecosystems, more of them are concluding that the interconnections they find among nonhuman organisms and the geophysical landscape connect them as scientists to those places as well. Jeffrey Alexander noted that Max Weber commented on the sharpness with which scientists perceive ethical issues: "The scientific, abstract way of thinking allows clarity about moral issues and provides a basis for critical skepticism toward the claims of worldly powers. If it is a true vocation, therefore, science, like politics, can fulfill a 'sense of responsibility,' and a teacher who can communicate such an ethic to his students 'stands in the service of "moral" forces.' "[25]

In their discovery of this new sense of responsibility and in their willingness to act upon it, conservation biologists and the movement their discipline represents constitute a break from the broader biology discipline and from particular subdisciplines such as fisheries. It is an entirely new way of viewing salmon, a new ideology and a new construction. Among the biologists who adopt a conservation biology perspective, the reality of what salmon are and what society's relationship to the salmon is and ought to be has dramatically changed. Self-reflexive biologist sociologists recognize this. For instance, Marc Raymond told me:

> For years this area of expertise was called "fisheries," which implies the harvesting of fish. In fact it's gone well beyond that, now. I think we're starting to understand that the fish, in and of themselves, have an inherent value. This certainly gets into value judgment and personal opinion as much as hard science, although I think there is some hard science discussion that can be held here as well in terms of what the role is that these fish play in structuring broader ecosystems. But there has been a change. There's been a very real change, and I'm not sure what's been the major driving factor there. But fish, wild fish in particular, are now viewed as being very valuable in and of themselves as self-sustaining populations far beyond their value on the dinner plate. That is a real change that's occurred, and it's one that continues to occur to the point where now we're looking at doing things like shutting down hatcheries, shutting down production facilities that for years and years formed the backbone of, or were the primary producer of, harvestable fish. But now, because they interfere with some of these wild populations, there's a real move to cut back production or even eliminate some of those facilities. I think it's just indicative of the change.

Conservation biologists construct a salmon that they identify with in the same sense that we all identify with certain human social groups, whether the group is a religious one, a political party, or a sports team. Key beliefs are always a part of such identifications, and for conservation biologists one of these beliefs is that humans are not the measure of all things. All those "things" are not somehow separated from us. Rather, salmon, to name one thing, matter in a profound way. It is they that have the historical claim to the Pacific Northwest. They coevolved with the forests, bears, streams, and canyons, not with industrial society, and industrial society should change its practices to respect this fact.

For biologists like Paul McGuire, the construction of salmon as inherently valuable and as intimately tied to their surroundings prompts a new way of understanding the fish. He views the inculcation of a new ethical perspective as especially important to the development of a conservation outlook. He observed:

> Most of our fishery biologists come out of the University of Washington. But up until recent times they haven't been taught the biological conservation ethic. I'm a graduate of that school. I talk from personal experience. You must have that conservation ethic, something I had to learn on my own. It must be encouraged and it must be put forth as the ethic to follow. We're in a stage now with our salmon populations where we should no longer be teaching the take ethic, the exploitation ethic. That should be fitted in only in particular situations. The major ethic that should be taught is the conservation ethic.

There should be no doubt how complete such a turnaround would be. What McGuire and others have in mind amounts to self-determination for salmon. Their ethics-driven approach calls for a substantial relinquishing of control over the fish: habitat improvements, severe alterations in dam operations, sharp curtailment of offshore and inshore fishing, and dramatic reductions—perhaps the complete elimination—of hatchery operations. The hatchery's homogenous salmon should be replaced by what existed for millennia: the heterogenous variability of wild salmon. McGuire argued these points forcefully, saying,

> I'd like to see far more appreciation for the salmon's welfare than we've given it in the past. . . . I'm a fisheries scientist, but I'm also a citizen of the Pacific Northwest, and it's my feeling that we've relied far too much on artificial salmon production and not enough

> on the natural production. We're at a critical stage in our time in the Pacific Northwest where we need to reassess our whole appreciation for the salmon. We need to get back to protection of natural processes if we really want to save the salmon.

Advocacy, Acceptance, and Resistance

Above, Reed Noss noted that biologists' concern for their research subjects rose as their perceptions of those subjects' well-being declined. Marc Raymond emphasized this point, saying,

> As we started to take a look at some of these biodiversity issues, which relate back to [Edward O.] Wilson's book on biodiversity back in the seventies, and as we began to understand that there were so many life forms disappearing so rapidly from the face of the earth, and as we began to understand the genetic diversity of the individual stocks here on the West Coast, I think all of that came together and generally led up to this new appreciation for the critters in the wild. I think that's been a large driver, the gradual recognition of their importance.

Fisheries biologists soon came to the conclusion that information regarding the loss of the fish, combined with their own growing ethical consciousness, was of little value if it was contained only within the profession. They needed an outlet, and they have taken an interest in affecting the public policy process. Conservation biology affords salmon biologists the sense that their work is relevant to society. As Alex Stand observed, "I think everybody wants to feel like their work is somehow relevant as opposed to purely an ivory tower exercise, I guess, as you're talking about people at universities. And I think the conservation biology field is appealing because we can all see the problems with the environment, the decline of plant and animal populations. It's an opportunity to apply some scientific rigor to those problems and hopefully contribute to solving some of them."

Some conservation biologists even see their on-the-job roles as that of activists. Terry Grey, a state agency fish manager and biologist, said, "What we do in my job, often, is we do act as advocates. We act as advocates for the fish. We have to if we're doing our job properly, because the mandate of my agency—and my own personal mandate—is to protect the resources that we're charged with managing. So we're very much advocates for the resource. Some people will carry that beyond." The "beyond" referred to advocacy outside of biologists' job descriptions, an emerging trend. For instance, in 1994 several Washington State Department of Fish and Wildlife employees filed ESA petitions on behalf of nine coho salmon runs.[26] One of them argued that most of those charged with advocating for the fish had abrogated their responsibility, saying,

> When you look at their record, there's eleven coho salmon stocks in Puget Sound that supposedly are managed for a natural production priority, which means the highest priority is to meet spawning escapement objectives. When you look at their record, the record for '91 is zero for eleven. Ninety-one is the brood year for the '94 run. These are all three-year-old fish. They won't tell that to you, or if they do they treat it as if low escapements is a natural phenomenon. These same fish that they failed zero out of eleven, some had fishing rates as high as 60 to 80 percent. When you know something about fish, you just go, "This is a goddamned con game."

It should not be surprising that such forceful, confrontational language has angered some traditional fisheries biologists. Advocacy almost always threatens the status quo and creates controversy. Conservation biology's biggest advocates are intensely aware of the divide between themselves and the old order. Marc Raymond said,

> It's a fundamental point of contention right now. There still is a very substantial proportion of the fish biologists here in the Pacific Northwest who do things that are related to actual production and harvest of the fish. It's related to exploitation of the resource, whether they are hatchery managers or involved in harvest management or involved in some aspect of sport fisheries. Those people very often are at odds with some of these individuals who've "got the religion," who've got this new perspective or have a different idea of what fish and fish biology is all about and how those re-

> sources ought to be managed. . . . So there's a lot of internal conflict as to how some of these things ought to be managed.

Bruce Alton observed,

> In fact, this boiled over into the national meeting last year, which was held in Portland. A substantial fraction of the business meeting of that year was devoted to the debate about whether we should rein in the advocates, as it were. My biased opinion is that those who were against advocacy had been burned by the advocates at some point along the way. Their programs had been criticized. There's bias on both sides. I think there have been a few times when the advocates have gotten a little ahead of the curve. Basically, I think without that sort [of] effort things would have been a lot worse off than they are.

The anxieties over advocacy have been the stuff of a running debate in the pages of *Fisheries*. An Oregon biologist wrote, "I disagree with our direction. . . . Serious advocacy is likely to transform AFS into a very different organization focused on conflicts and having mostly nonprofessional members. . . . Why? Advocates need conflicts, must be powerful to be effective, and must be effective to survive."[27] The editor replied that a survey of AFS members showed 92 percent wanted to see the Society increase its efforts at affecting environmental policies. This portends nothing less than what Max Weber called conflicts over the "non-bureaucratic top"[28] of organizations, which includes both the personnel there and an organization's guiding normative principles. At stake is the future of the profession, the future direction of fish biology, and quite likely the future of salmon.

Perhaps the strongest counterargument to the conservation biology perspective, more compelling even than concerns over self-interest (for example, that the profession might lose its power, become ineffective, and wither), was expressed by an academic biologist who frequently publishes on issues relating to fish farming:

> One of the things that came out of the American Fisheries Society—there's been a lot of literature in the last couple of years in *Fisheries*—was a draft statement that said that economics should never play a

> role in decisions like this. If you're a purist, that's great. But economics is going to play a role. To assume otherwise is unrealistic. . . . I feel uncomfortable when we take people out of the equation. In some of these things, that's what they're trying to do, take man and reality out of the equation.

Recall that fish farming was *the* driving interest among the founders of the AFS. Feelings such as these may represent the last gasp of a dying ichthyological ideology. On the other hand, they may also represent the tenacity of the much broader and more compelling rationalization ideology.

COMMONALITIES WITH THE FISHERIES PERSPECTIVE

As substantial a break with the past as the conservation biology approach to salmon biology represents, the fracture is not complete. In salmon biology, clear indications of the anthropocentric, utilitarian ethic so instrumental to the creation of the fisheries field and so dominant for so long continue to be found. Will Perry, who coauthored an ESA petition for coho salmon, told me, "Every fish stock out there is just like an annuity account: if you manage it correctly, you get a continuous stream of benefits for this generation and the future. If you don't, if you start pulling the principle out, it isn't going to work. It isn't going to be renewable. None of this is a surprise to anybody who knows anything about fish." "Annuities," "benefits," and even "renewability" reflect a traditional, use-oriented perspective. Even Paul McGuire, the most ardent of the conservation biologists, said something similar: "We need to begin to gain a greater awareness of how to repair the habitat and get some of this habitat back into natural production for these natural fish to use. We have many, many miles of stream in this state that are underseeded; the habitat is there but there are no fish to use it." For him, "production" referred both to increasing wild population numbers for their own sake and, over the long term, for human economic gain, or at least for a stable, sustained economic take.

Perhaps even more indicative of the tenacity of the dominant fisheries perspective than this continued embrace of the utilitarian attitude among those who profess to change it is that conservation biologists argue that their work and their outlook must be grounded in "good science." Stephen Fogarty contended:

> Yes, you can't help but get involved. But on the other hand, if you get emotionally involved with your species and you show that, you have less value as an advocate for that species. Everything has to be empirically based. It has to have data that's well developed. As soon as you break away from what your data actually shows and go into a broader thing, you may in fact destroy your whole credibility.
>
> *Q. Is it possible to develop an emotional attachment to a species and be biologically fair and sound?*
>
> FOGARTY. Oh, yeah. I think so.

Paul McGuire was equally adamant, saying that he was willing to take on the advocate's role so

> long as it is scientifically-based. I never want to get into the position where I am seen as an advocate that is not scientifically based. As long as I have the data, I'll let the data do the talking. The data speaks. I'm a strong believer in that. If the data points in one direction, I believe in going with it. If it means stepping forward beyond the scientific literature and making a public statement on behalf of this resource, I have no problem with that as long as I have the data backing me. I don't want to get out on a limb without the data support behind me. It's those people who see the writing on the wall, and the data is there, who are not willing to either be convinced or to step out and admit that we need to do something that I have a problem with.

In this perspective much value remains in rationalizing Nature through science. The issues must be calculable and predictable. Advocacy and emotion supported by data are sound and sensible. Without scientific backing they are dangerous.

It is important to note that ecocentrism's valuing of nonhuman entities is a different sort of valuing, one that does not rely on data. It is a cultural and philosophically based perspective that is "irrational" in light of the dominant worldview and that even resists science as it is traditionally conceived. Max Weber's concept of the "irrational" is a conceptually rich one that has at least two meanings. The first—though not the meaning of immediate interest to us here—might best be termed *organizational irrationality*. It reflects Weber's understanding that irrationalities are integral, even indispensable, to the activity of institutions. No organization can actually pursue policies of efficiency, control, calculability, predictability, and productivity in a pure, ideal sense; contrary behaviors—irrationalities—may be necessary to keep employees or customers happy, or they may be practiced out of tradition, or for any number of other reasons. Let employees take an extra fifteen minutes for lunch on Friday or make hamburgers by hand, not by machine—the examples of deliberately irrational behavior are legion. And in practice, pure rationality is unachievable.

The second meaning of irrational, which I will call *reasoned irrationality*, is a lot more fascinating, and it is highly relevant to conservation biologists' behaviors. Where, precisely, did rationality come from? Was it given by some unseen maker of social laws? Of course not. Powerful social forces—the "logic of capitalism," the "invisible hand," call them what you will: they are entirely social—decree that efficiency, control, and the rest are the only rational ways to organize our lives. Seen from this viewpoint, rationality suddenly loses its near-biblical status and becomes a mere product of society. In summarizing Weber's viewpoint, one author put it pithily: "Reason itself could not found reason."[29] Incredibly, *rationality was founded upon irrationality*. Values, beliefs, attitudes, and norms are at the root of rationality. Society declares certain of these to be reasoned, logical bases upon which to found an economic—and social—system, but it could have declared that others suit the purpose. That is precisely what many other societies through time have done. In a sense, our notion of rationality was chosen (although the reality was that it took centuries for its "logic" to be worked out—further evidence of the irrationality of it all), and we might have selected other values instead.

Because any rational system is founded upon irrational value preferences, it is possible that conservation biology's "irrational" valuing of salmon based upon their inherent worth might some day be seen as rational. The pseudophilosophical approach known as "deep ecology" espouses just this type of alternative rationality. Embracing an ecocentric perspective, deep ecology argues both that "good science" is indispensable to ecocentrism and that ecology undermines industrial society's rational edifice. Wrote one deep ecologist, "The subversive insight of ecology is that not only is everything connected with everything else but there is a literal intermingling of person and Other, of mind-in-nature."[30] Conservation biology is simultaneously biological and ethical in orientation. It is not clear what the impact of this combination will be on either it or the broader discipline, although the opportunity for considerable cognitive dissonance appears to be at hand.

Given all their conflicts and commonalities, it is reasonable to expect that advocates of both the fisheries and the conservation biology perspectives will persist in their disagreements. The biologists left me with little doubt on this score. To exemplify it, I offer the opinions of two biologists about the efforts to save the first salmon run listed as "endangered," the Redfish Lake (Idaho) sockeye. A biologist who identified with the conservation biology perspective said,

> I guess I feel that we have to draw a line in the sand somewhere in terms of how far we're going to let the environment and the salmon population slide. I think if you don't draw it there I suppose the next step is to say, "Well, we aren't going to have salmon in the Snake River, period—let the chinook go extinct, too." It seems to me that this is just as good a place to draw the line as any. And in fact the expensive steps, which are going to be the steps to improve the survival of the young salmon going up to the ocean down the Snake and Columbia Rivers, that will benefit the sockeye are also going to benefit the chinook, which are still more abundant and more widespread than the sockeye.
>
> I guess you could probably make the argument, and I don't know that I would be eager to make that argument, either, that if the chinook weren't also present in the area, that if it was just down to that one sockeye population, maybe your money would be better invested somewhere else. On the other hand, given that the most expensive parts of your recovery effort, which are going to be the passage issues, are also going to benefit the chinook, I think it makes a lot of sense to go ahead and try to save both of them. That would be my argument.

Another biologist argued from the classical fisheries perspective, saying,

> To my mind that's such a ridiculous thing to try to maintain a sockeye salmon run in Idaho. I can't think of a worse place. What a travesty or whatever! What a waste of money to try to produce fish there when we have sockeye coming out our ears [in places other than Idaho]. . . . My feeling is, and I'm all for salmon, but if there were two fish left in that lake, I'd kill them both and get it over with rather than spend millions of dollars to try to build something in the worst place in the world to try to raise sockeye salmon. . . .
>
> If you're in Idaho and you want salmon, my feeling is, pay for it. Have your salmon. That's no problem for me. But when you come and ask me to pay higher utility rates to pay for your salmon. . . . The problem the industry has right now, especially fishermen, is that there are too many fish. Fishermen today, not here but in Alaska, are fishing harder, working harder, and making less money because they have to catch more to make less money because the price is so low because of oversupply.

Both views maintain a keen awareness of economics. Rationalizing salmon persists, so strong is the historical construction. Yet conservation biologists seek to strike a balance between ecology and economics. Through their interpretation of ecology they appeal "to a common human self-interest in its maintenance. . . . And it is in this sense that change is a question not just of ecological sermons, ethics, and 'moral oughts,' but of ecological self-interest."[31] For conservation biologists, what is right *ecologically* turns out to be right humanistically—for humans. This represents a startling break with traditional fisheries, where the motto might have been: What is right *economically* is right humanistically.

CONCLUSION: BACK TO THE FUTURE

The road to conservation biology within salmon biology has not been an easy one, and the journey is far from over. Conservation biology represents an assertion of self-determination and of freedom on the part of biologists in the face of a society and a discipline driven by control and power. Moreover, conservation biology constructs salmon as important in themselves, and this is a bombshell of a break for fisheries biology. Some in the fisheries profession appear to be fearful of the impact that the trend toward a conservation biology orientation could have, especially on the ideal of scientific objectivity. They argue that without objectivity, science loses its political legitimacy. Conservation biology-related themes, especially advocacy, have been at the center of a running debate in the fisheries profession. That advocacy and activism have caused such a stir within the fisheries community is perhaps to be expected.

Yet these issues are not resolved even within conservation biology generally. When *Conservation Biology* editor Reed Noss published several articles detailing the damaging impacts of cattle grazing on public lands, headed by his own antigrazing editorial in a 1994 issue of the journal, it drew a sharp response from three respected conservation biologists. They wrote that their peers should resist Noss's call for an alliance with antigrazing activists "because of our deep concern about the issue of scientific objectivity," and they concluded by warning that "such thinly veiled advocacy seriously threatens the Society's mission to be the nation's and the world's respected source of responsible scientific expertise in conservation matters."[32] As with the debates in the fisheries community, at issue was political credibility, legitimacy, and status. Advocacy—even when supported by data—threatens to transform biologists into *mere* advocates, the implication being that advocacy science will be easily dismissed by decision makers as partisan.

Ironically, given the controversy over advocacy, and advocacy on behalf of salmon in particular, activism was of utmost importance to the earliest members of the American Fisheries Society. The official AFS history notes, "The new organization's members were activists. Within a few months of formation, the AFCA lodged protests with the State Department and British Canada against obstructions in the St. Lawrence River which impeded the migration of salmon. Prompt promise of law enforcement, removal of obstructions, and prevention of illegal fishing was forthcoming."[33] Fish biologists were activists for fish and for the profession from the start. Today they are merely going back to the future.

What the conservation biology perspective represents is an effort to simultaneously free salmon biologists and salmon to pursue self-determination. Conservation biologists, still constrained by rationalizing economic and governmental forces, are seeking to convince their colleagues that they should be able to say what they will regarding the objects of their studies. Some of them wish to become advocates, to petition for the protection of salmon under the Endangered Species Act and to speak out publicly on behalf of the fish. The goal of these control/power-relinquishing acts is to free the salmon to continue living, evolving, and playing their role as an important, even "keystone" (essential) species in the Pacific Northwest. And one senses that the biologists are being freed at a personal level. They are experiencing a bit of "ecstasy," to use a term from Peter Berger's wonderful book *Invitation to Sociology,* that exhilarating, anxious act of "stepping outside . . . the taken-for-granted routines of society," or, in this case, the profession.[34]

The practical, impersonal effect of activism like this is to challenge the forces of rationality that have dominated fisheries biology since its inception. Max Weber wrote, "The Puritan wanted to work in a calling; we are forced to do so."[35] Like Emile Durkheim before him, Weber saw "voluntarism"—freedom and self-determination—slipping away before rationality's strangling influence. Salmon biologists' emerging perspective represents a counter to that trend.

As they pursue this self-determined thrust, these biologists also are constructing a salmon that is radically different from those which industrial societies have created over the last century. Granted that the fish continue to have economic meaning and import, relinquishing control over them has the effect of fundamentally altering their political meaning. Politically and economically, the fish continue to be important. However, as symbols they represent not a potentially endless resource, but rather they are simultaneously fragile and resilient fellow denizens of this planet. In the conservation biology construction, human interactions with the salmon change from utilitarian and use-based to ecocentric and instructive of the tensions between humans and Nature.

Paul Ehrlich has argued for "a much less anthropocentric, more egalitarian world, with greater emphasis on empathy and less on scientific rationality."[36] Conservation biology represents a new "soft science," one that rejects the old standards of objectivity and economic ends and replaces them with advocacy and process. If anything saves the salmon, such a construction of Nature is likely to be the key.

7 Salmon Wars and the "Nature" of Politics

THROUGHOUT THIS BOOK I have discussed the ways that microlevel social interaction combines with macrolevel social institutions to produce biologists' varied constructions of salmon. In this chapter I emphasize the macro-level exclusively, with special emphasis on the so-called Salmon War between the United States and Canada. This exploration of the Salmon War will see us join the social construction of Nature with the social construction of communities. Two important lessons emerge in the process. The first is that social constructions really can be changed. The second is that only by consciously creating, and vigorously pursuing the realization of, our preferred constructions of Nature can those changes come to fruition expeditiously.

We will get to the details of the Salmon War shortly. First, let us examine some observations by those who studied a similar struggle: the demise of eastern Canada's great fishery. Then we will trace the chronology of the Salmon War, and we will conclude with an analysis of the social constructions of salmon that have emerged from the conflict, as well as their import for society's relationship with Nature.

POWER TO THE PEOPLE?

It was no later than the 1500s that Europeans began fishing off of what are today Newfoundland and Nova Scotia. Their takes were substantial, and the fishery was relatively stable for centuries. It was only in fairly recent times that changes in fishing rules, fishing technologies, and the accompanying industrialization of the fishery imperiled the fish, and so the fishery, to the point where it collapsed beginning in the early 1970s. There simply were no more cod or turbot to catch.

In one of the first sociological articles on the social construction of Nature, Ralph Matthews examined the impact of the collapse on fishery management's construction of the Canadian east coast fishery. Matthews noted that those constructions shifted dramatically as the crisis loomed. Prior to the fall of the fishery, policy was "biologically driven." But more recently the fishery—those who are doing the fishing and the areas where they fish—has been "viewed . . . as open access 'common property' " that was shared but unregulated by laws or even locally derived norms; this new construction emerged despite the fact that there was "no evidence to support such a conception."[1]

Based upon his research and his reading of history, Matthews concluded that "the way in which we undertake to regulate and manage our natural resources rests on our 'social construction' of them as having either their own rights or as instrumental to our needs."[2] This dichotomy may not be as inclusive as Matthews asserts (for instance, a combination of the two constructions may be possible), but there are clear parallels between his conclusion and those that I drew above regarding the split in salmon biology. Fishery biologists, especially those emphasizing "applied" research needs, approach their task instrumentally, driven by what benefits might obtain from the use of salmon: the "economic rent," as one biologist put it, that may be extracted from them. Conservation biologists, in contrast, understand the salmon as possessing "inherent rights," values akin to the freedoms enshrined in the U.S. Constitution. We do not look at other persons merely as economic entities, and this "ecocentric" perspective embraces a similar approach to other animal species.

As he sorts through the policy makers' constructions, Matthews distinguishes between three constructions of fisheries. One is the "social" fishery, which, in several variants, gives primacy to cultural values, community, and lifestyle, sometimes in combination with economics. A second primary construction is more narrowly economic and instrumental, emphasizing the efficiency and "maximum economic yield" of fisheries with little or no regard to the culture of a fishery. This is a distinctly rationalized construction—maximum economic yield is nothing but productivity, predictability, calculability, and control rolled into a single phrase that implies that "the economically desirable fishery is one which harvests the greatest economic *value* with the least expenditure of economic effort."[3]

Matthews' third fishery policy construction is especially relevant to the analysis in this chapter. It is a combination of the social and economic fishery policy constructions, but like the others it fails to add biology to the calculation. It is commonly referred to as "sustainable development," or "development that meets the needs of the present without compromising the ability of future generations to meet their own needs."[4] One writer described sustainable development as "equity and conservation" rolled into one, and those very concepts are at the heart of the Pacific Salmon Treaty discussed below.[5] Equity stresses how a resource will be divided amongst those that exploit it, and "conservation" concerns limitations on that exploitation.

Sustainability, then, is less about biology than it is "an economic and a social strategy," wrote Matthews.[6] Where, in such a calculus, is biology to be found? There is no mention of, for instance, salmon as valuable in and of themselves or as possessing species rights, only mention of salmon in reference to human ends. In his conclusion, Matthews wrote that " 'sustainability' is about interests and not altruism, and about the needs of [society] rather than about the needs of nature. Until that is clearly recognized and management systems are developed to regulate the interests involved, then the fishery—like the rest of nature—remains endangered."[7] Thus understood, the sustainability construction actually rejects an ecocentric perspective, and in the process it risks the very Nature that is aspires to preserve.

The sustainability perspective also places much of the norm-setting power in the hands of the state. Matthews' research suggests that too much centralization of power is not the solution, however. He favors giving a substantial amount of real decision-making power to those communities whose livelihoods depend on their ability to maintain long-term, sustainable relationships with entities like salmon. He writes that "those who fish from these communities must be given real power to decide on allowable catch allocations, appropriate use of gear, licensing regulations, and a variety of other dimensions of fishing practice. In other words, they must have more than a simple advisory role."[8] It is telling, then, that in the chronology of the Salmon War that follows, the fishers have been largely excluded from direct discussions with one another in an atmosphere where they can make decisions for themselves and bridge the state-established policies that separate them. The United States and Canadian governments have created their own constructions of the Pacific salmon fishery. In the process, they have rejected the notion of a devolution of power to the level of the fishing community.

ANATOMY OF A FISH WAR

In 1977, nations around the globe agreed that "territorial" waters—the ocean areas that nations could claim as their own property, and therefore could control in essentially the same ways that they control terrestrial areas—would extend two hundred miles from shore. So long as fish swim only within a single nation's waters, the new treaty posed no problem. But salmons' peregrinations are far-flung. At sometime in their lives, many of the salmon that leave the rivers of the United States and southern British Columbia as smolts travel through waters off of Oregon, Washington, British Columbia, and Alaska, in addition to spending lengthy periods in international waters. Thinking of these vast wanderings, Matthews wrote, "Fish are a fugitive resource. That is, they are capable of moving themselves and, in order to harvest them, they must be captured."[9]

For a few years there was turmoil in the salmon-fishing communities of the West Coast. The most pressing question was, Whose salmon are these? The Law of the Sea Conventions elevated ownership of the oceans' "resources" above all else, and conflicts were created out of whole cloth once the treaty took effect. States, provinces, tribes, and the federal governments of both the United States and Canada have laws to promote production of wild salmon, and they also operate large numbers of fish hatcheries. Because of this, wild and hatchery salmon alike were seen as respective states', provinces', or nations' property. In this view they were less fugitives than they were homing pigeons. Because the Law of the Sea Conventions recognized only nations, not jurisdictional subdivisions like states or provinces, any resolution of the question of ownership would have to come through a further treaty, this one between the United States and Canada.

CAPTURING A FUGITIVE WITH A TREATY

An effort to do precisely that had been underway for seven years before the Law of the Sea Conventions were finalized, but it would be as many more years still before the two nations could agree to the terms that would govern how their commercial fleets would fish for the lucrative salmon. When the Pacific Salmon Treaty was finally signed in

1985, the ceremony included the nations of Canada and the United States, the states of Alaska, Oregon, and Washington, the province of British Columbia, and Native American tribal representatives. The involvement of players other than the two nations is fascinating, since under the U.S. Constitution, only the federal government can negotiate treaties. As it turned out, the participation of nonfederal groups would prove to be a major stumbling block to future dispute resolution. On the U.S. side, the states were the real players in the dispute.

The Pacific Salmon Treaty "is based on two principles: conservation and equity," according to the Canadian Department of Foreign Affairs and International Trade:

> The conservation principle obliges each country to prevent overfishing and provide for optimum production. The equity principle provides that each country should receive benefits equal to the production of salmon originating in its waters. It means that no country should catch more than it produces. These two principles are linked: conservation requires both countries to protect and rebuild stocks; equity gives a framework for allocating catches.[10]

These principles embody one of the central confrontations in society's relationship with Nature: that between ecology ("conservation") and economy ("equity" of "production"). One observer quoted above blithely praised sustainable development for seamlessly welding conservation and equity. In theory, perhaps. But rarely if ever do the two combine with such ease in practice.

We often see this tension played out in the media, where journalists pit environmental and corporate interests against one another. Each side seems to represent a definite social construction, but the picture is frequently muddier than such a facile analysis would have us believe.[11] On occasion, political figures attempt to close the apparent chasm that divides these interests by declaring, as did Bill Clinton and Al Gore, that we can have a healthy economy and a healthy environment both. Yet the conflict remains because, I think, groups' social constructions are never part of the formalized environmental dispute-resolution process. Decision making takes place at a shallower level where the "social" is reduced to economics, or at best to attitudes.

For instance, in order for wolves to be reintroduced to Yellowstone National Park, the U.S. government developed an environmental impact statement, as directed by federal law. It set out a limited set of "scenarios" for reintroduction, one of which was selected after public hearings, changed a bit to reflect public sentiment, and became "the" reintroduction plan. Yet at no point along the way did anyone attempt to seriously consider the "social impact" of the wolf reintroduction. Attention to groups' social constructions of the wolves might have identified some of the potential rips in the social fabric that resulted just from the hearing process, not to mention the reintroduction itself. Doing so could help avoid disputes in the first place.

Most often when we hear politicians trying to bridge the ecology-economy gap, they are talking about clean-air legislation or the like. We can have smokestack industries, they say, without the smoke. We can keep jobs and corporate profits (and the sundry ancillary benefits, such as donations to political campaigns) and still have a cleaner environment than would have been the case had not corporations been forced by law to clean up their emissions. A similar argument was made by President Clinton with regard to the northern spotted owl: that we could save the ancient forests where the owl lives and still have plenty of logging and milling jobs to go around.

But unlike with clean air or the spotted owl, with something like salmon we are confronted by an ecological entity that is itself economically valuable, the animal equivalent of a redwood tree or a vein of gold, and that only intensifies the conflict. Clean air cannot be priced very easily (the new fad of "oxygen bars" notwithstanding), and spotted owl probably isn't very tasty (though anti-owl bumper stickers intimate that they "taste just like chicken"). In contrast, humans attribute to salmon both ecological *and* economic worth. So, is it possible to have enough fish in the water and in our supermarkets? Optimistically, the parties to the Pacific Salmon Treaty said that it is.

Almost immediately after its formation, the Pacific Salmon Commission, established to carry out the treaty's directives, ran into a series of roadblocks. Not only did ecology and economy have to be balanced, but the interests of two nations, four states and provinces, and the Native American community had to be taken into account as well. It took four years just to agree on the data to be used in allocating equity: which nation gets how many fish. Once that was accomplished, the actual negotiations over equity could proceed. But they didn't. An intractable dispute arose over whose fish were whose. Fish that originate in Alaska, British Columbia, Washington, and Oregon all swim through multiple jurisdictions. So Alaskans can catch "Canadian" salmon off of Alaska, and Canadians can catch "Washington" fish in the waters of British Columbia.

The treaty's equity principle directed that each *nation* could catch fish in proportion to what it "produces" (production, of course, being the number of fish from hatcheries and from wild spawning). But accusations arose that fishers from some areas were taking more than their share or that, in pursuit of equity, conservation was being hindered. For the Canadians, this meant that their decision not to dam the Fraser and other rivers to preserve wild salmon runs—especially the lucrative sockeye stocks—was for naught because Alaskans were overfishing the sockeye as well as coho and the big chinook, or "king," salmon. When they heard this, Alaskans responded by arguing that the real problem is that Canada manages its own fishing industry poorly. For Washington, Oregon, and Indians from the United States, the concern was that the many endangered runs—especially of coho salmon, but really of just about all of the species—could literally be wiped out in one dip of a seiner's huge nets. For their part, the Canadians claim to have cut back on their coho harvest. In the treaty's terms, the Canadians are anxious about equity, while the Americans—at least those south of the border—emphasize conservation.

The dispute dragged on from 1989 until 1993, when Vice President Gore promised to work steadfastly to resolve the issue. He did not succeed. In 1995, Canada proposed mediation to end the impasse. The United States agreed, and in 1996 Christopher Beebe, a New Zealand diplomat with decades of experience, was appointed to help the sides resolve the dispute. Eventually, Beebe walked out on the talks because the U.S. government would not budge. (When Beebe's secret report was leaked in 1997, it strongly favored the Canadian position that it deserved compensation from the United States because of the latter's history of overfishing Canada's salmon, though Beebe acknowledged that violations of the letter and the spirit of the treaty had occurred on both sides.)[12] There followed intensive discussions by high-level governmental representatives and assurances by governors that an agreement would be reached.

Instead, the dispute began to heat up. The Canadians refused to participate in developing a plan to catch salmon during the 1996 season, and the United States responded by creating its own. More negotiations followed, this time among representatives of the states and British Columbia. After some hopeful signs, another failure. In the summer of 1997, Canada proposed binding arbitration to bring closure to the Salmon Wars, but the United States would have nothing to do with it. Yet more high-level representatives were appointed, and so the story progressed (or didn't).

THE SALMON WAR GETS HOT

Until mid-1997 the Salmon War was a diplomat's dream: not really a war but with plenty of conflict, lots of brinkmanship, an array of confusing issues, and lots of internecine feuds thrown in for good measure. Repeated unresolvable disputes were nearly overcome, only to see the entire effort fall back into the abyss, as when—just as the sides were near an agreement—a U.S. representative revealed that she had not been given the authority to make any agreements. The Salmon War had everything short of a "hot" war, that tragic moment when even the Kissingers leave the stage and the Schwartzkopfs take over. For instance, in 1994 Canada levied a "transit license fee" of C$1,500 on all U.S. fishing vessels passing through Canadian waters, a policy that would effectively put many fishers out of business. The stated purpose was to make Americans pay for taking Canadian salmon, though in reality the Canadians were groping for a way to pressure a resolution to the Salmon Treaty impasse. Vice President Gore again intervened, promising to "get to yes," and for a time things cooled down.

But in the summer of 1997, things got really hot. In May, Canada began detaining U.S. fishing boats for violating a law that required them to inform the authorities that they were using Canadian waters. Even vessels having nothing to do with the dispute were caught up in skirmishes. In July dozens of Canadian fishing boats blockaded an Alaska-based passenger ferry in the port of Prince Rupert, B.C., and there were reports of gunshots exchanged between U.S. and Canadian fishing boats.[13] Then Glen Clark, the premier of British Columbia, announced that he planned to end an agreement that allowed U.S. submarines to use Nanoose Bay off Vancouver Island as a torpedo testing range. Clark declared, "It's a full-scale fish war."[14] Amidst all of the tension, there was a lighter side. In May 1998 five fishermen stripped to their skivvies, à la *The Full Monty*, in the hope that they might move stalled negotiations along.[15] All this over salmon.

There are several underlying problems that helped bring tensions to the boiling point. Foremost is that the treaty did not include any firm dispute-resolution guidelines. When there is a falling out over the Pacific Salmon Treaty, no set authority exists to resolve the issues. And, as alluded to above, the treaty is atypical because it includes parties other than just the nation-states. Quite often the Canadian federal government and British Columbia clash "over the way the west coast fishery is managed and how negotiations should proceed with the United States."[16] On the U.S. side, each of the three voting members to the Pacific Salmon Commission—Alaska, Washington, and the Indian tribes—has a veto over any proposal, and their interests often conflict. Alaska is generally seen as the worst of the lot because its fishers tend to capture more salmon heading for Canadian spawning grounds than any other fleet—nearly as many as the Canadians catch themselves.

Even when there are occasional successes—as occurred in 1998 when the Canadians and Washington state agreed on a sockeye salmon management plan (no doubt owing to the terrifying specter of another fisher *Full Monty*)—it only exacerbates tensions. B.C. premier Clark said his federal government was on its knees when it negotiated the sockeye agreement.[17]

Clark also objected to a June 1999 revision of the Pacific Salmon Treaty. He posed " 'two fundamental questions.' First, does the treat 'meet the conservation test of protecting endangered B.C. coho salmon?' And second, does the deal 'ensure a fair sharing of non-endangered salmon that are surplus to our conservation needs?' Clark said no on both counts, pointing out that last year, while B.C. fishers caught no Canadian coho, Alaskans landed 800,000; and that this year, Canadians will catch none of their coho, while Alaskans will be fishing."[18] Clark was so upset that neither he nor any other representative of the British Columbia government appeared at the treaty signing ceremony. The treaty revisions, which seemed to include cutbacks in the permissible catches of some runs, were welcomed by others, primarily groups concerned about conserving United States salmon runs and improving opportunities for commercial fishers in Oregon and Washington. But

British Columbia's lack of support likely means that the revisions will hold little weight.

CONSTRUCTING COMPLETE COMMUNITIES

One thing is relatively certain: Without an agreement between the United States and Canada over how to manage the humans, salmon will be "overfished," to use the popular euphemism. Not all salmon runs will be exterminated, but many more will be unless the nations and the states can reach an understanding. And the longer the process drags out, more and more human communities are, like the salmon, becoming endangered. The current atmosphere is one where there is tremendous social and economic pressure to fish to the ragged edge and beyond. It is a place where the future of salmon and those who fish for them take a backseat to politicians and international competition. In the process, neither the salmon nor fishing communities really seem to matter to those making the decisions. Fish and fishers are constructed by their absence in the negotiations. Federal, provincial, and state governments are the only ones at the treaty table. They claim to represent "fishing interests," but prior works make me question just how benevolent they will be.

In his study of the Canadian east coast fishery's demise, David Ralph Matthews wrote, "If there is one lesson to be learned from this book, it is that there is a strong historical and practical basis for allocating the regulation of fishery property rights to the local fishing community."[19] Ultimately, in the Pacific salmon example, what needs to be balanced are the interests of the salmon, the technologies used to exploit the salmon, and our laws and theories that emphasize property rights as applicable to a species like these fish. Matthews says that true power must be vested in local communities. They need self-determination regarding how to fish, when to fish, and how many fish to catch. He so argues because he recognizes that the fishers understand the connection between their community and the salmon community: that they are, for all intents and purposes, one community. Perhaps if the nation-states acknowledged this and gave up just a little control, the people who live this society-salmon connection could resolve the dispute themselves.

However, powerful and controlling interests stand in the way of such a radical reconstruction of the human-salmon relationship. In the midst of the current Pacific salmon struggle, one Canadian official was quoted as saying, "There is very real competition among all those groups to be sure they don't get less than their share."[20] So perhaps it is a system that rewards economic competition and formal rationality that is at the root of the problem. Even if some of these communities wanted to give new meaning and new reality to their relationship to the salmon, they could not do so on their own and realistically hope to survive for long, given our society's emphasis on efficiency, productivity, predictability, and control. This is so because others would be waiting to take their place.

So there is a place for institutions. Not only must they be willing to accede to some degree of self-determination for fishing communities, they also must be willing to safeguard those freedoms by creating new forms of economic behavior by vesting control over markets to those who fish and thus who are most affected by the markets. And most important, only when governments that typically ensure economic interests and values over all others decide that *they* are willing to re-construct the human-salmon relationship as an ecological one rather than an economic one will the true salmon wars, the wars between society and the salmon, ever be over.

Like Matthews, Raymond A. Rogers, who fished commercially in Nova Scotia for twelve years, also wrote about the collapse of the Canadian east coast fishery. He observed that, in such reformulated constructions as the one Matthews advocates, "there is little difference between 'nature' and the 'social.' "[21] It is too late to save the east coast fishery, and it is nearing that time in the Pacific. Rogers writes that we need to set aside our social construction of self and of Nature, which "has all but destroyed the other aspects of natural identity as the processes of market economy strengthened in the modern period."[22] In place of this narrow, exclusively human-focused meaning of Nature, Rogers asks us to consider creating one that allows for individuality (and thus personal creativity, freedom, and self-determination) within the context of a greater whole (ecosystems) and that re-creates our sense of community to include humans, the places in which they live, and the nonhuman denizens of those places. Rogers and Matthews have witnessed what can happen when the interests of fish and fishers are overlooked by those whose social constructions of society and Nature exclude community.

Certainly this deliberate physical and cognitive reconstruction of a culture's relationship with Nature sounds utopian, incomplete, and uncertain. Few of us have yet to attempt it, and some will always prefer an approach that would construct Nature by eliminating it. Senator Slade Gorton of Washington State, always one to embrace rationality over principle, said, "There is a cost beyond which you just have to say very regrettably we have to let species or sub-species go extinct."[23] Others, fortunately, are seeking a genuine reconstruction of the human-salmon tie. John A. Fraser, formerly Canada's ambassador for the environment, said, "This is about these magnificent fish and all of us who have the tendency to destroy, but also the capacity to protect and conserve. No one owns these fish, even less does any particular interest group. This resource is held in trust by all Canadians for each succeeding generation of our peoples."[24]

The rhetoric is lofty as it is laudable. It will be the actions that matter—for the nations, the fishing communities, and for the salmon. Matthews wrote that "some clear mechanism must be found whereby these various *competing* interests can either be met or adjudicated. Unless that happens, and soon, each country will continue to protect its own interests in much the same manner that each individual under the supposed conditions of common property was motivated to exploit the resource in his or her own individual interest."[25] Can Slade Gorton's single-minded narrow-mindedness be replaced by John Fraser's expansive view of what and who really matters? If so, it will indicate that the old rationality macroconstruction has lost its grip on North American society just a bit and that perhaps society has the courage to reconsider its place in the world. Perhaps we, too, are Natural.

CONCLUSION: NATURE AS WE WANT IT TO BE

At stake in the struggle between Canada and the United States are ways of life—parallel ways of life, in many respects—not just individual fortunes or national pride. The lives of those who fish the seas for salmon are inextricably bound to the animals that they fish for. These self-employed fishers are the humans who will suffer should the salmon be overexploited, and their suffering will be of the deepest kind. Not only are they facing the loss of income, but of a way of life. What is at issue is something that is shared by a comparatively small number of us, but we all can sympathize with these persons because so many of us have lost that something, too: community.

I do not mean to overly romanticize community—surely it can be oppressive and produce ignorance when it becomes insular. But in industrial times community has been transformed into society: a place that is hardly a place at all, a world where we often are cut off from sources of love, support, caring, mutual understanding, and sustenance. For those in fishing communities, the salmon make that all possible. In a subtle way, the fish are constructed by these communities as enabling these human communities to exist. Do we have the political will to save salmon for the good of these communities and for the good of the salmon as well?

Raymond A. Rogers wrote that "conservation issues have to be in the perspectives of participants who have collectively recognized they are part of both human and natural communities, at all times. This kind of integration at the local level offers some hope of long-term conservation of the fish stocks, although how it will be instituted in the context of an increasingly deregulated global playing field is difficult to say."[26] These sentiments mirror Matthews's, who urged policy makers to vest local fishing communities with control, and in the process to reconstruct fisheries and fishing.

Rogers adds another, crucial stipulation: "Modern culture has to consider the appropriateness of seeing the natural world as a standing reserve from which the huge industrial apparatus can expect a return on its investment."[27] Salmon, ancient forests, beautiful canyonlands, soaring mountains: none will be safe so long as society's dominant construction of Nature is a utilitarian one, so long as they all are constructed as dollar bills with fins, dollar bills with branches, and so forth. We can never see Nature for what it is, only for what we want it to be. The power of social transformation—of identifying a new social construction of Nature and bringing it about—is in our hands, communally. The outcome will depend on the pressure that we bring to bear on our political leaders, the impact that we have through the organizations we join, and the changes that we make in our individual lives.

8 Constructing Nature—and Experiencing It

I SET OUT on this project with the hope that, by studying salmon and salmon biologists, we might be able to come to a better understanding of how other "natural" entities, and perhaps even Nature itself, are made real and given meaning in people's lives. This is a basic and necessary endeavor if we are to study the relationships between society and the Environment, yet until recently this effort has been almost totally neglected. In this spirit, the first section of this chapter summarizes the concepts that can give us some guideposts to comprehending the social construction of Nature.

In the concluding section I pause to acknowledge the thing that sociologists eschew above all else: the individual. I know from my own experiences—and from discussions with many others—that an unconstructed *nature* is out there, and it needs to be acknowledged. To us individually, those close encounters with nature unmediated by culture and society are important. Eventually, however, we must come out of the forest or up from the depths and live social lives. When we do that, and when we attempt to convey our raw, experiential interactions with nature, the meaning-filled social world comes flooding in. Suddenly, we are not just individuals feeling a sense of connection to nature. To the contrary, we are members of interconnected, embedded communities, interest groups, economic classes, races, genders, and organizations. Our social identities and their corresponding interests, outlooks, and biases begin to color and shape even directly experienced nature. Socially, at least, Nature never simply *is*.

Toward a Sociology of Social-Natural Interactions

We now have the methodological and conceptual/theoretical guideposts necessary to examine how different specific nonhuman entities, such as salmon or redwood trees, and the generalized nonhuman other known as "Nature," are socially constructed. The method suggested here emphasizes the importance of a historical and comparative approach, and the concepts developed in this book that are transferable to other instances of social-Natural interaction include the control/power, self-determination/freedom spectrum; macrosocial structure; science/technology; and cognitive/physical factors.

Methods. "Historical" and "comparative" methods involve either taking a long-term perspective that usually relies on historical documents or juxtaposing different social groups (or societies). Regardless of the approach used, the idea is to develop better insights by not relying on a snapshot of time or a single social setting for one's data. Both methods highlight constructed social reality, enabling researchers operating in the constructivist tradition to understand the social processes that result in changed meanings. Historical and comparative perspectives compel researchers to account for differences in what otherwise might be treated as the same phenomenon. For example, although many commonalities can be found among the primary comparison groups used in this study—U.S., Canadian, and tribal biologists—the distinctive social settings enabled us to better understand the range of constructions of salmon that this social group holds—including the fact that it is less and less a single social group at all. Finally, for sociologists meanings are uninteresting phenomena unless they are examined as malleable constructs. Historical and comparative analysis facilitates the discovery of the changing, evolving processes that effect new meanings through time and across cultures.

Control/power and self-determination/freedom. The most important of the theoretical findings emerging from this study is that social groups' constructions of nonhuman entities like salmon, and of Nature generally, may be located on a spectrum between the poles of control and power, on one hand, and self-determination and freedom, on the other. On the control/power side is found the ability to direct others' behaviors, while the self-determination/freedom end of the spectrum denotes the lack of fetters on a group's actions and outlooks. As I developed these concepts in Chapter Six, social groups construct their behavior and beliefs as seeking self-determination/freedom against opposing groups that possess some amount of control/power. This study, and my subsequent research on the social construction of wolves, leads me to believe that whenever there is conflict over a Natural entity, this tension is present.

Macrosocial structure. Social structure's impact on constructions of Nature cannot be overemphasized. Berger and Luckmann noted the crucial role of the structural embeddedness of individual and small group activities in the creation of meanings. In any society, macrosocial structure acts in two ways upon social constructions. It enframes new meanings, and in time some of those meanings become "legitimated"—part of the shared taken-for-granteds. Macrosocial structure also affects *which* meanings become dominant, and thus are taken for granted. The agents of rationalization in capitalist industrial societies that are especially important in shaping the meanings of Nature are the economic, political, governmental, and technological/scientific forces.

Those forces imbue the Environment with a utilitarian meaning—it is society's to control, manipulate, and use. However, what is especially telling are the methods employed to rationalize Nature, particularly what constitutes the *efficient* use of Nature. Here, governmental logic becomes indistinguishable from economic logic. For example, a public agency advocates for the construction of a dam while promising that a unique run of salmon will be preserved for all time in a nearby government-operated hatchery. The agencies involved are likely to argue that the river and the salmon will "operate" more efficiently for economic purposes—they will produce more electricity, more fish, and more social benefits than the undammed river generated. Rationally, it is wasteful to do anything else. A wild river is an inefficient river.

Science and technology. These forces bear special mention in regards to the rationalization of Nature because *science and technology are the instruments through which rationality physically constructs Nature.* Science is drafted by governmental and economic interests to more efficiently use the Environment. And, in this time when science challenges religion as the dominant mythological belief system, science is not only used to make Nature more efficient, but to justify those rationalizations as well. Salmon biologists, for example, are indispensable in governmental and corporate efforts to establish limits on the amount of fish that can be caught by commercial and recreational anglers. Even when "conservation" is the goal of rationalizing forces, their influence may be seen. For example, biologists involved in efforts to "recover"—bring back from the edge of extinction—the Snake River sockeye salmon openly speak of the monetary costs of various options; efficiency remains of utmost importance even when "everything possible" is being done to save Nature.

Cognitive and physical constructions. In these pages I also have introduced the distinction between cognitive and physical constructions of Nature. Although some constructivists argue that reality exists only in the mind, my approach accepts that material things are the objects of constructions. Those physical phenomena—salmon and the hatcheries that they often are raised in, for example, or the tags or transmitters salmon carry to allow their tracking—are where social constructions of Nature emerge in their starkest, least ambiguous form. We can see what is occurring in a hatchery, and we can witness biologists tagging salmon. Certainly, these structures and behaviors must be interpreted—meanings are as crucial here as they are in the purely cognitive aspects of constructions—yet they are distinct from language because their existence is substantive and not completely abstract and symbolic. Quite simply, the fact that there is something material being constructed is important. Objects and behaviors are more easily problematized and struggled over than are ideas and concepts alone (even the struggles over religious "faiths" and secular "ideologies" usually have a physical or behavioral component to them that is central to the conflict, such as the performance of rites or the exercise of rights).

As for the cognitive factors contributing to social constructions, the concern is *not* a society's attitudes, values, beliefs, and norms, but rather the meanings attributed to phenomena. Meanings may or may not reflect attitudes and beliefs. The former are complete and reflective of persons' realities, those taken-for-granted aspects of our lives that we see as permanent and act in accordance with, while the latter are often fleeting and are sanctioned by social structure. For example, persons may have the attitude that recycling is important, but they do not recycle on a regular basis; recycling has not been made a part of their ideology, of who they are: their social reality.

Part of my basis for asserting this distinction between attitudes and meanings/realities is the method used to gather data. Meaning is the product of qualitative research and of the substantial amounts of time that are spent with research participants, whereas attitudes and the like are usually examined using surveys that do not entail direct contact between researchers and participants. There is little or no effort to develop relationships and understandings of participants' lives, and this is essential if meanings and lived realities are to be identified and fully explored (details of the method used here may be found in the Appendix).

Knowing a Meaningless Nature

George Herbert Mead, one of the great sociologists of all time, argued that anything that cannot be communicated between humans by definition has no meaning, because meaning is a social phenomenon.[1] In concluding this discussion about Nature's meaning, I would like to briefly explore a way of knowing that is *meaningless,* that is intimate and personal, not social. Through these meaningless human-Nature interactions, Nature is known as *nature,* unproblematic because it is ours individually and no one else's. I take this brief excursion to clarify the realism that underlies my view of Nature.

A small group of writers, and large numbers of less published persons, have experienced nonhuman entities in meaningless ways—through strange and awe-inspiring encounters that cannot be communicated to other humans without losing the intense reality of the experience. It is ironic, and perhaps futile, to attempt to write of such things, both because the subject is intensely personal and because in the act of communicating the knowledge and filling it with meaning, the knowledge—raw *experience,* really, the most certain of all knowledge—is lost.

Henry David Thoreau, for one, claimed to experience such moments. He wrote, "The mind that perceives clearly any natural beauty is in that instant withdrawn from human society."[2] We become a part of something other than the human, social world when we see with the intensity that Thoreau spoke of. This withdrawal from society is also the subject of Annie Dillard's brief essay, "Living Like Weasels." One afternoon Dillard strolls to a neighborhood pond to escape the rat race. As she sits on a stump, she turns while following a bird's flight and, in an infinite instant, her eyes lock with those of a weasel looking up at her from beneath a wild rose bush.

> Our look was as if two lovers, or deadly enemies, met unexpectedly on an overgrown path when each had been thinking of something else: a clearing blow to the gut. It was also a bright blow to the brain, or a sudden beating of brains, with all the charge and intimate grate of rubbed

> balloons. It emptied our lungs. It felled the forest, moved the fields, and drained the pond; the world dismantled and tumbled into that black hole of eyes.
>
> . . . I tell you I've been in that weasel's brain for sixty seconds, and he was in mine. Brains are private places, muttering through unique and secret tapes—but the weasel and I both plugged into another tape simultaneously, for a sweet and shocking time. Can I help it if it was a blank?[3]

Something broke the spell, and Dillard's mind was "suddenly full of data;"[4] she found herself groping about trying to comprehend it all, but it all was gone.

A student of mine once wrote of a similar experience. A doe had wandered into his camp during a solo backpacking trip. It was a wilderness area, and my student had no reason to think that the deer was tame. Yet she returned several times over the two days my student was at the site. They seemed to communicate through momentary glances and subtle gestures. He struggled to give voice to the experience, to give it meaning, but it was so different from anything he had ever experienced, he said. It was as if the wall between him and the deer no longer existed. Finally, he gave up and resigned himself to the thrill, or perhaps the shock, of what had happened.

Contemplating such stories, Neil Evernden wrote, "We can certainly know the concept *nature;* as a container, it is ours completely. But the contents can never be known as encountered in experience if we begin with a *denial of experience.*" Once we accept the notion that unmediated, meaningless experience is *the* means of understanding nature, we can begin the process of understanding anew "by asking ourselves what it would mean to actually encounter the world *for the first time.*"[5] The meaningless moments that Thoreau, Dillard, Evernden, and others write about are times when we see with the blank mind of a weasel or a child, when we close in an instant the millennia of distance between us and the other that was so painstakingly constructed by philosophers, historians, sociologists, industrialists, educators, and parents.

Yet the very act of attempting to convey what went on while the eyes were locked, while the experience was at its richest, thrusts one back to the social world. For language is but a symbol, a reification and an abstraction of experience and the experienced. It is never the experience itself. Those who peer into the weasel's eyes undergo something extraordinary. The experience takes place while staring at a stunning sky or while lost in reverie by a peaceful pond. The moment is one of completion, as one entity complements another. It is real, the most real thing one can do, something akin to seeing one's self or seeing inside the soul of someone you hold dear. We are not supposed to be able to see ourselves, Erving Goffmann says. We do not live in a society where such self-seeing is possible; certainly it is not commonplace. And because of this, its possibility has been discounted.

However, I am equally convinced that voicing the experience, even acting on it, brings us into the domain of the social. Dillard's essay constructed a weasel and her experience of him; it gave meaning to that interaction. What came before the writing does not seem, to many sociologists, to be of sociological interest. Mead would call it meaningless, disparaging it at least in terms of sociological import, if not in a more colloquial sense. But are such experiences meaningless? To the contrary, they are, I think, important to sociology in and of themselves. That they are part of the human experience tells us that a heretofore unappreciated phenomenon is occurring, and that alone is worthy of our curiosity, even if it does not appear "social" initially. Moreover, those who experience such intersubjective complementarity appear compelled to act upon those experiences. Dillard acted through the written word, my student through attention to his daily activities and their impacts on Nature. How and why do they act?

Paradoxically, through words and deeds meaning is given to these "meaningless" activities. Social constructions are created through these simple acts because writing and speaking reduce the personal and the private to terms that are socially apprehensible, available for our everyday, commonplace interpretation and understanding. Roland Barthes wrote of such things, saying,

> Mythical speech is made of a material which has *already* been worked on so as to make it suitable for communication: it is because all the materials of myth (whether pictorial or written) presuppose a signifying consciousness, that one can reason without them while discounting their substance. This substance is not unimportant: pictures, to be sure, are more imperative than writing, they impose meaning at one stroke, without analyzing or diluting it. But this is no longer a constitutive difference. Pictures become a kind of writing as soon as they are meaningful: like writing, they call for a *lexis*.[6]

The same may be said of dance or of sculpture. Through mythical, social "speech," regardless of its medium, the sublime becomes the mundane. Meanings are created and are shared. And this is not a bad thing, for in the sharing new opportunities for new meanings, new ideas, and new actions arise. This is the stuff of social change.

John Muir once wrote that if he could just get people to the forest they would understand what he had experienced and would love Nature as he did. Muir saw that *experience* is the key to understanding human-environment interactions. One of the numbing features of the rationalized lives that we lead is that we so seldom have the time, money, or energy to pause at the side of a pond and to stare into the weasel's eyes. Simultaneously—and not unconnected, I think—we claim that our lives have no meaning. Perhaps Michael Crichton was correct when he wrote that we are encountering the limits of science. We have come to the realization that science is not Nature, nor even a reflection of it. After four thousand years, perhaps Western society is at the moment when we recognize that it is *we* who are reflected in Nature. Fortunately, enough of *nature* remains that we can continue to imbue it with new meanings and to search for ourselves even as we search for it.

APPENDIX
METHODS AND RELATED LITERATURE

This appendix includes two sections. The first, on methods, is intended for those interested in how the data for this study were gathered and analyzed. The second section discusses the social construction of science and technology literature, prior research in the social construction of Nature, and critics of the constructivist perspective within environmental sociology.

DATA GATHERING AND ANALYSIS FOR THIS STUDY

The observations in this book are the product of nearly three years of research and analysis. Data gathering took various forms and occurred in a variety of locales. I interviewed twenty-three salmon biologists and one other wildlife biologist from the United States, Canada, and Native American tribes, and I spoke with numerous others in noninterview encounters. Those interviewed included ten federal agency employees (five Canadians, all of whom worked at a large government research station, and five from the United States, including three salmon hatchery biologists and two agency researchers), nine university professors (one of whom was Canadian), three state agency employees, an American forestry corporation biologist, and one private consultant, a Canadian. I conducted fieldwork at two salmon biology conferences—one for regional members of the American Fisheries Society that brought together American and Canadian biologists, the other for American tribal biologists—and I attended numerous public hearings on salmon-related topics where salmon biologists were present. My other field experiences included watching biologists at work in fish hatcheries, accompanying others as they gathered data on rivers and in other settings, and observing biologists over several months in a laboratory that conducted DNA research on salmon.

My initial interview participants were selected both because of convenience—they worked near my university—and because several of them were highly respected in the salmon biology community, judging from their publication records, which I checked using *Science Citations Index* and a major volume reviewing the research on Pacific salmon[1] prior to the interviews. The interviews were tape-recorded and lasted an average of seventy-five minutes each. In each I asked biologists whom else I should speak with for my study, and from the resulting list I developed my next round of participants, a process known as "snowball sampling."[2] Most of the initial participants were fisheries biologists. My later discussions with conservation biologists were a result of serendipity. I happened to be at a conference session where a biologist was speaking in terms reminiscent of conservation biology. I interviewed that biologist and developed a snowball sample of like-minded persons from there.

All of the interviews were semischeduled—I always had a list of questions to ask, though I asked other, unplanned questions as well, and the interviews that I felt were most productive were more like conversations than question-and-answer sessions. New topics arose in all but the final interviews, and in no case were two participants asked the same questions. This openness to shifting research emphases and exploring new topics reflects my "grounded theory" analytical approach, explained below.

In no sense were any of the participants or other sources of data for this study generated randomly, which is a hallmark of "quantitative" social research. The intent of this project was not to generate "scientifically" (statistically) valid and reliable results. The goal was more modest: to create concepts that might aid in the *understanding and interpretation* of human-nonhuman interactions, a venerable approach in sociology and anthropology. Having said this, it is important to note that the comparative method that is essential to grounded theory does much to address concerns about validity and reliability. Researchers actively seek counterexamples to strengthen their theoretical ideas. In addition, my reading of publications in the fisheries community, and my interactions with fisheries biologists who were present at presentations of my findings on two occasions, leads me to believe that my interpretations accurately reflect biologists' lifeworlds. More important than any evidence of validity or reliability, smiles and nodding heads among scientists told me that I had interpreted their profession in ways that resonated with them.

GROUNDED THEORY

Grounded theory is one of several popular approaches to inductive theory building and data analysis in the social sciences.[3] As is the case in most inductive research, the grounded theorist usually does not know where the research will take her or him. Instead, concepts of interest are teased out of the data, such as interviews or field notes. Those ideas are "tested"—explored for their importance—by gathering more data, and the process ends when no new ideas are developed.

A substantial amount of literature discussing grounded theory is available.[4] Barney G. Glaser and Anselm L. Strauss coined the term, by which they meant "the discovery of theory from data systematically obtained from social research."[5] That is, concepts are directly tied to empirical observations. This approach stands in contrast to deductive—or, less charitably, "armchair"[6]—theorizing, in which researchers create theories prior to collecting data.

Some argue that grounded theory has a positivistic (rigorously scientific, truth-seeking) and quantitative feel to it (for instance, labels for concepts are called "codes"—which is what data are called in quantitative research—and the reliability and validity issues mentioned above preoccupied the method's founders). For these very reasons my friends Patti and Peter Adler have called grounded theory "the darling of mainstream sociology."[7] My own approach has been to treat grounded theory as thoroughly qualitative. Validity and reliability issues do not concern me in the least, but the details of working effectively with data to obtain an understanding of my participants' worlds do. As a guide for the systematic analysis of interview and observational data, grounded theory is unsurpassed.

Kathy Charmaz has led the way in making grounded theory useable to scholars in diverse fields. Although her interpretation is finely nuanced, she essentially writes of four basic steps to grounded theory.

First is the gathering of "rich" data—usually interviews and observations. Charmaz asserts (rightly, I think) that the researcher is part of this process, not some unseen, "unbiased," disinterested data recorder. Instead, she says that "the interaction between the researcher and the researched *produces* the data."[8] The researcher is there in the act of knowledge production and is not separable from that process. Recognizing this, researchers become *self-reflexive,* aware that they are part of the social scene that they are examining and that they affect what is going on there. Indicative of this is the use of the first person in their writings, as I have done throughout this book. By writing in the first person I declare that this text has an author, one with thoughts, emotions, and preconceptions that may affect what you read, much as I try to be fair to my research participants' viewpoints. But more importantly, I have been deeply affected by what the other participants in this process have told me. They initially allowed me into their worlds. But in time I became seduced by it. I began to see salmon a little like they did. No ethnographer can honestly claim to be part of a culture other than his or her own. But we try nonetheless to get as close to our hosts' lifeworlds as we can.

Second, the data are constantly analyzed. That is, even as data are being gathered, researchers interpret them and attempt to tease out theoretical nuance. For example, in the midst of an interview ideas that potentially may give additional understanding are either tried out on a participant—"It just occurred to me that. . . . Does that make sense to you?"—or are jotted-down in a researcher's notebook for possible later investigation. The analytical process is intensified when field notes or tape-recorded interviews are transcribed. The researcher examines the data closely, initially relying largely upon participants' own words or raw field observations for the conceptual labels (the "codes" mentioned above) that reveal the meaning of the topic in question.

Third, the data gathering and analysis steps are repeated several times. The insights from the immediately preceding round of data analysis are "tested" by gathering and interpreting yet more data. Theoretical ideas are always examined in the "real" world through interviews, observations, and the like; researchers treat their data as their only touchstone, and every insight has to be supported by the data. This strengthens the inductive theory that is generated through the research because the theoretical concepts are being challenged each step along the way.

The upshot of this challenging approach is twofold. First, topics that initially appeared to be disparate and unconnected are, with each iteration, drawn together under ever more general conceptual labels to yield abstract theoretical concepts. Also, many of the early codes taken directly from participants' words are set aside as these conceptual labels are identified. Second, some topics that seemed promising during one phase of the research process will be dropped later. This may happen for several reasons: their explanatory power may prove insufficient, for example, or they may not be relevant to later respondents during interviews. In sum, the testing of ideas and concepts hones and intensifies the data gathering, because fewer and fewer theoretical notions are seen as relevant with each passing iteration.

Finally, when the data yield no new "theoretical" insights, no new ideas, the data-analysis portion of the research endeavor is completed. The final step in the process, writing the study's findings, is undertaken with an eye to tying together multiple strands of thought while remaining faithful to research participants' worlds. The general theoretical findings are typically very few, but they include a host of subconcepts, all of which closely reflect the data. The final product is "data intensive" as well. Quotes from interviews and field notes are provided to support the conclusions and to enable readers to share the ethnographer's intense exposure to the social group that was studied.

THE INTELLECTUAL HERITAGE: PRIOR WORKS

Two constructivist literatures are especially relevant to this study. The first is the social construction of science and technology, which is one of the more controversial applications of the constructivist approach. Works by several scholars writing in this vein were cited in the preceding chapters, and here I address some of the more problematic of their assertions. The second set of literature comes from the newer but rapidly growing studies of the social construction of Nature.

SOCIALLY CONSTRUCTING SCIENCE AND TECHNOLOGY

Even as "social constructivism" has become the rage in sociology, Berger and Luckmann's "classical" approach has been largely ignored, with some peril I think. An apt example of the peril emerges from one of the most active areas of scholarship in social constructivism, the social construction of science and technology (SCOST). Both the classical approach and SCOST share core concerns. Like Berger and Luckmann before them, SCOST scholars such as Bruno Latour, Wiebe Bijker, Trevor Pinch, and others,[9] emphasize meaning creation as inherently problematical. Two of the foremost SCOST advocates put it simply: "Scientific findings are open to more than one interpretation. This shifts the focus for the explanation of scientific developments from the natural world to the social world."[10] This is essentially an application of Berger and Luckmann's notion of the relativity of knowledge and reality.

However, distinctions between these constructivisms emerge when considering some of the finer details of the dominant approach in SCOST, known as the "strong program" or "strict" constructivism.[11] For example, the strong program is guided by an idealist philosophical belief that "reality" is a mental construct. In contrast, Berger and Luckmann did not question the existence of material reality. While they were not immune to philosophical concerns, Berger and Luckmann appear to have recognized that, as sociologists, ontology and epistemology were not central to their enterprise. They wrote that "the subjective experience of everyday life" that is at the core of phenomenological perspectives like theirs "refrains from any causal or genetic hypotheses, as well as from assertions about the ontological status of the phenomena analyzed."[12] Rather, what was crucial were the social processes that give material reality its meanings.

A second distinction emerges from the strong constructivists' philosophical idealism, for they easily slip into an epistemological quagmire wherein their accounts of science are constructed as truthful and stable, yet science's accounts of Nature have no such philosophically solid ground upon which to stand. For instance, Pinch and Bijker wrote that "in investigating the causes of beliefs, sociologists should be impartial to the truth or falsity of the beliefs, and . . . such beliefs should be explained symmetrically."[13] Somehow, social constructions identified by scholars are supposed to be objective ("impartial") and create unbiased ("symmetrical") analyses. In so arguing, strong constructivism deceives itself, for constructions of constructions nevertheless are constructions. As Stephan Fuchs wrote, constructivists should reject such "discredited metaphors of realist epistemologies" and admit that their studies "are simply self-exemplifying;"[14] constructivist analyses, as constructions themselves, possess no unique claim to truth. Accepting this simultaneously dismisses angst about the validity of one's worldview and takes constructivism to heart.[15]

Berger and Luckmann's more modest approach never asserted the vaunted status that strong constructivists seek. Theirs was a *sociology* of knowledge. The classical approach provides social analysts with an insightful critical perspective from which to work, one that does not rest upon philosophical concepts that have been debated through the ages. Moreover, theirs is a general framework for analysis, not a road map: a paradigm instead of a theory. Berger and Luckmann encourage researchers to consider historical context and to appreciate both interactional and macrolevel forces as they examine the emergence and maintenance of meaning. Beyond that, the specifics are left to the researcher.

SOCIALLY CONSTRUCTING NATURE

Although the "social construction" concept probably originated in sociology, until recently the social construction of Nature was entirely left to other disciplines to explore.[16] A measure of how far behind others environmental sociologists are may be found in William Cronon's major edited volume of interdisciplinary perspectives on the social construction of Nature;[17] it includes no chapters by sociologists. Perhaps this is because the initial reaction by some respected sociologists to the social construction of Nature was quite negative,[18] a topic addressed below. Nevertheless, other sociologists have developed a range of theoretical approaches,[19] some of which I will comment on here, and at least one policy study.[20]

Catton and Dunlap: The First Social Constructivists of Nature

William R. Catton Jr. and Riley E. Dunlap are considered by many to be the cofounders of the environmental sociology subdiscipline. At the very least, they were instrumental in placing environmental sociology on the conceptual radar screen of the broader discipline. In the late 1970s and early 1980s, Catton and Dunlap challenged their colleagues to acknowledge the importance of the Environment to society. They argued that the time was ripe for a Kuhnian paradigm shift in sociology, writing, "Disciplinary traditions and assumptions that evolved during the age of exuberant growth imbued sociology with a worldview or paradigm which impedes recognition of the social significance of current ecological realities. Thus, sociology stands in need of a fundamental alteration in its disciplinary paradigm."[21] Their goal was to make explicit the irrational (in Weber's terms) underpinnings in the " 'Human Exemptionalism Paradigm' implicit in traditional sociological thought, and to develop an alternative 'New Ecological Paradigm' which may better serve the field in a post-exuberant age."[22] The upshot of their argument was that sociology had constructed social systems that were "exempt" from the forces of Nature.[23]

In this, Catton and Dunlap saw what might be termed an *undernaturalized* approach to sociology. As a discipline founded in a time of growth and plenty, sociology seemed to have no need for Nature. Nature was conquerable—controllable at the very least—and was merely the clay from which societies were built, nothing more. Catton and Dunlap recognized that *sociology had constructed a Nature by default,* by omission, and that construction reflected the dominant social paradigm. Nature did not exist in sociology any more than it was a pressing issue in society.

Catton and Dunlap first called for a sociology of the Environment amidst the shock of the oil embargoes of the 1970s, not long after Americans awoke to the Environment as a social issue. They warned of the precariousness of our species' future, with Catton imploring his colleagues: "Our existing sociological paradigm . . . obstructs clear comprehension of the kind of 'sea change' that has affected our future. . . . Despite old domain assumptions that have caused us to define such matters as outside the boundaries of sociology, understanding this sea change in the human condition is now, in my judgment, *the* most fundamental challenge for sociologists."[24] But the sociological seas have not shifted as Catton and Dunlap envisioned, although in 1993 they wrote that "recent years have seen a dramatic resurgence of interest in the field and at least the beginnings of its intellectual revitalization."[25]

Landscaping Nature

Part of this "intellectual revitalization" is found in the social construction of Nature literature, and the first to attempt a systematic constructivist approach were Thomas Greider and Lorraine Garkovich. They drew upon Berger and Luckmann's sociology of knowledge as the foundation of their argument that imbuing Nature with meaning is a social process. Any particular social construction of Nature is that group's "landscape:" "the symbolic environment created by a human act of conferring meaning on nature and the environment."[26] By examining the meaning-filled actions of humans as they interact with the nonhuman, it becomes clear that "cultural groups socially construct landscapes as reflections of themselves."[27] "Nature" says as much about us as we can say about it.

Greider and Garkovich also observed that power—or "control" in the terms used here—is a primary determinant of a society's dominant constructions. They wrote that the ability to manipulate meanings implies that dominant social constructions "of nature and the environment get imposed, sometimes through the use of force, on others with less power."[28] These manipulations of meaning imply that change of some sort is being created. They wrote that "what is important in any consideration of environmental change is the meaning of the change for those cultural groups that have incorporated that aspect of the physical environment into their definition of themselves."[29] Environmental change implies new physical and cognitive constructions, and, when social groups feel strong affinity for the land or some aspect of Nature (such as the connections between fishers and salmon), those groups' self-understandings are at stake as well.

This further implies the presence of confrontation surrounding Nature's meaning, a topic addressed by other sociologists. For example, in Jan E. Dizard's *Going Wild: Hunting, Animal Rights, and the Contested Meaning of Nature,* conflicting social constructions of Nature emerge out of the tense interactions between two social groups in opposition to one another.[30] The comparative and ethnographic approach used by Dizard, also present in Greider and Garkovich's article and in this study, has emerged as the primary methodological tool employed by researchers in this area.

Other Understandings

Lines of inquiry similar to those pursued by environmental sociologists have emerged in the discipline but outside of environmental sociology proper. This work by symbolic interactionists and constructivist theorists has anticipated the general "social construction of Nature" by exploring how pets,[31] specific species of animals,[32] animals generally,[33] and mushrooms[34] are treated as symbols and are given meaning by owners, experimenters, viewers, pickers, and others. Arnold Arluke and Clinton R. Sanders contrasted this literature to that which predominates in sociology, writing, "Most sociological research is anthropocentric (or human-centered) and focuses on relationships among humans,"[35] and this is as true for environment-oriented sociologists as for any others. In much of the environmental sociology research, animals, plants, and inorganic entities like canyons and streams are the objects of social attitudes and behaviors, but it is rare that they are treated *ecocentrically,* as subjects that can be examined to reveal deeper insights about the roles of particular social forces in persons', animals', and places' realities.

THE ANTICONSTRUCTIVISTS

A Change of Face

In an intellectual irony, Catton and Dunlap, the first constructivists, have rejected the more recent constructivist turn. In a recent paper they detail five supposed shortcomings in the social construction of Nature literature. Although they specifically critique several works by Frederick Buttel and his colleagues on the social constructions of "global environmental change" (GEC) such as the "greenhouse effect,"[36] their comments reflect a general anxiety over the notion that Nature could actually be constructed and that science's viewpoints could be social as well as factual products. Following a paraphrasing of each of Catton and Dunlap's points, I offer a brief rejoinder:

1. *The constructivist view obscures society's role in the physical creation of Environmental problems.* Other than social groups, what force creates Environmental problems? The real issue is our appreciation for both physical and cognitive realities, because physical problems have no meaning absent our constructions of them. As I argued in Chapter Seven, social constructivism gives insight into the conflicting meanings that underlie Environmental issues, including the struggles between dominant and less powerful constructions for primacy and

legitimation. No analysis of Environmental problems is complete without an appreciation of these constructions.

2. *The lack of firm scientific data regarding GEC is an inadequate reason to argue that GEC is socially constructed.* The lack of data means that "qualitative," intuitive approaches to interpreting climate change are necessitated. In such uncertain conditions, science is compelled by society to create its own myths. These topics were explored in Chapters Three, Five, and Six. Regarding scientific claims more generally, the approach taken here does question whether science is as objective as it is made out to be in textbooks, and it inquires about the purity of the knowledge science produces as well. Powerful, controlling social groups often are "in" the science. However, as noted in Chapter Five, it is possible to both maintain a skeptical view toward scientific practice and to accept its demonstrable insights.

3. *Constructivist approaches of the sort employed by Buttel reduce GEC to a claims-making debate that ignores real Environmental issues.* Constructivism is more than an exploration of social problems, with differing sides lining up to do battle (some practitioners' contrary approaches notwithstanding). It examines the social bases of the claims being made and the social and ecological impacts of those claims, as I have tried to show throughout this book.

4. *The ultimate effect on environmental sociology of a constructivist approach would be to ignore human-nonhuman interaction because it overemphasizes the social forces affecting Environmental claims-making while ignoring the ecological.* This would probably be a correct assertion if strong constructivism were the only variant, but implicit in my approach is that nonhuman entities are important, too. They are subjects and not mere background, secondary objects. Constructions have as their foci physically extant human and nonhuman things. What we make of them is precisely what this investigation has sought to understand.

5. *Constructivism promotes human exemptionalism of the sort that Catton and Dunlap argued was so harmful to Nature, and ultimately to humans.*[37] This is similar to a concern that fellow environmentalists have shared with me: that constructivism is really a tool of devel-

> opers and destroyers. First, such a stance implies that there is one true Nature, yet I am confident that a constructivist analysis would demonstrate that environmental groups do not share a common understanding of Nature even among themselves. Second, I think this approach can be a powerful tool for demonstrating the underlying assumptions of those who would have humanity continue to lord over the earth as thoughtlessly as industrial society has. And it may have the potential for educating people about their ecologically destructive taken-for-granteds.

Loren Lutzenhiser offered an explanation for realists' anxieties when confronted by constructivism, writing that "just as traditional sociological self-understandings are uneasy with 'technical' and 'biological' topics, we can now add emergent interpretivist perspectives that see natural environments largely as social constructions—*nature* as a potentially important social variable risks becoming *mere nature* as socially variable."[38] I have great appreciation for the ecocentric morality implicit in Catton and Dunlap's critique of constructivism. They seem to be saying, "We have the data. Humanity is harming the planet in the worst of ways. And just as we were making some headway in getting people to understand this, here come the constructivists." My response is that constructivists attempt to probe deeply into the world of those who advocate one or another position in society, and some semblance of impartiality (however impossible in practice) is important.

Nothing in constructivism dictates that interested persons should do nothing based upon their feelings, attitudes, and understandings of issues and their interpretations of the data. But constructivism does seek to demonstrate that the control and power to make Nature is within our hands and that the Nature that is being made is a product of society, not of the thing itself. The real issue for environmental sociology at this moment should be how the Environment and Nature are given meaning, not the correctness of the sides squaring off in debates over supposed "Environmental problems" (debates that themselves exemplify conflicting meanings of Nature). We need to address Nature sociologically. As I asked in the first chapter, Which social forces appear best situated to influence meaning-creating processes? What interests are served by doing so? And how do less powerful social actors effect changes in meaning in the face of dominant, hegemonic social forces? Revealing the roles played by sociologically significant actors in society's constructions of Nature would make it possible to examine those actors' motives. Combined with data from ecological studies, scholars could develop more complex and more complete analyses of the impacts of certain behaviors on the Environment. That is the promise of the social construction of Nature.

Murphy's Failed Critique of Constructivism

Aligned solidly with Catton and Dunlap is Raymond Murphy. In his book *Rationality and Nature,* Murphy attempts a savage indictment of the social constructivist approach.[39] He presents an incomplete summary of Berger and Luckmann's work, particularly a brief section in their book that mentions Nature, but most disturbing is that at no point does Murphy treat "nature" or "the natural environment" as problematical in the least. Ignoring Nature's historically and culturally contextualized meanings enables Murphy to use the term uncritically, as if there really were a universal, unassailable, real Nature out there. Lacking a comparative or historical perspective, Murphy appears insensitive to the variety of constructions of Nature that historians, geographers, anthropologists, and, increasingly, sociologists have identified. Environmental sociology needs to problematize its core concepts—Environment and Nature—if it is to act with authority in rectifying the abuse being heaped on the planet by industrial society.

NOTES

CHAPTER 1. NATURE IN THE MAKING

1. Michael Crichton, *Jurassic Park* (New York: Ballantine Books, 1990), pp. 312–13.

2. C. S. Lewis, *Studies in Words* (New York: Cambridge University Press, 1967), p. 37.

3. I. G. Simmons, *Interpreting Nature: Cultural Constructions of the Environment* (New York: Routledge, 1993).

4. See: Bruno Latour, *Science in Action* (Cambridge: Harvard University Press, 1987); Harriet Zuckerman, "The Sociology of Science," in *Handbook of Sociology*, ed. Neil J. Smelser (Newbury Park, Calif.: Sage, 1988), pp. 511–74.

5. Rik Scarce, *Eco-Warriors: Understanding the Radical Environmental Movement* (Chicago: Noble Press, 1990).

6. Tim Palmer, *The Snake River: Window to the West* (Washington, D.C.: Island Press, 1991), p. 5.

7. Anthony Netboy, *The Columbia River Salmon and Steelhead Trout, Their Fight for Survival* (Seattle: University of Washington Press, 1980), p. 3.

8. Peter L. Berger and Thomas Luckmann, *The Social Construction of Reality* (New York: Anchor, 1966).

9. Loren Lutzenhiser, "Sociology, Energy and Interdisciplinary Environmental Science," *American Sociologist* 25 (1994): 58–79.

10. See William R. Catton Jr., *Overshoot, the Ecological Basis for Revolutionary Change* (Urbana: University of Illinois Press, 1980); and idem, "Homo Colossus and the Technological Turn-Around," *Sociological Spectrum* 6 (1986): 121–47.

11. Stephan Fuchs, *The Professional Quest for Truth: A Social Theory of Science and Knowledge* (Albany: State University of New York Press, 1992), p. 194.

12. William Cronon, "Ecological Prophecies," in *Major Problems in American Environmental History*, ed. Carolyn Merchant (Lexington, Mass.: D. C. Heath and Company, 1993), p. 10 (emphasis in original).

13. Neil Evernden, *The Social Creation of Nature* (Baltimore: Johns Hopkins University Press, 1992), pp. 96–97 (emphasis and capitalization in original).

14. Will Durant, *The Story of Philosophy* (New York: Simon and Schuster, 1933), p. 400.

15. Netboy, *Columbia River Salmon and Steelhead Trout*, p. 16.

16. Berger and Luckmann, *Social Construction of Reality*, p. 89 (emphasis in original).

17. David Ralph Matthews, " 'Constructing' Fisheries Management: A Values Perspective," *Dalhousie Law Journal* 18, no. 1 (1995): 48.

18. See: Jeffrey C. Alexander, *The Classical Attempt at Theoretical Synthesis: Max Weber*, vol. 3 of *Theoretical Logic in Sociology* (Berkeley: University of California Press, 1983); Donald N. Levine, "Rationality and Freedom: Weber and Beyond," *Sociological Inquiry* 51, no. 1 (1981): 5–25; and Ann Swidler, "The Concept of Rationality in the Work of Max Weber," *Sociological Inquiry* 43, no. 1 (1973): 35–42.

19. George Ritzer, *The McDonaldization of Society: An Investigation into the Changing Character of Contemporary Life* (Thousand Oaks, Calif.: Pine Forge, 1996).

20. Max Weber, *The Protestant Ethic and the Spirit of Capitalism* (New York: Scribner's, 1958), p. 24.

21. Ibid., p. 181.

22. As noted above, formal rationality is means-oriented and emphasizes the anticipation of outcomes and the calculation of consequences. In contrast, Weber's "value rationality" is distinguished by "a belief in the intrinsic value of the action, regardless of the consequences, and is oriented to a set of values" (Raymond Murphy, *Rationality and Nature: A Sociological Inquiry into a Changing Relationship* [Boulder, Colo.: Westview, 1994], p. 30). Murphy insightfully notes that the formal rationality-value rationality dichotomy reflects the extremes of the debate over the Environment in North America: Nature as an instrument, controlled to serve human ends, or Natural entities as valuable in and of themselves without valuing them according to whether human gain might be realized by their manipulation or even their preservation (such as for aesthetic reasons). See also Chapter 6; Bill Devall and George Sessions, *Deep Ecology* (Layton, Utah: Peregrine Smith, 1985); Bill Devall, *Simple in Means, Rich in Ends: Practicing Deep Ecology* (Salt Lake City, Utah: Gibbs Smith, 1988); Roderick Nash, *The Rights of Nature* (Madison: University of Wisconsin Press, 1989); Holmes Rolston III, *Environmental Ethics* (Philadelphia: Temple University Press, 1988); and idem, *Philosophy Gone Wild* (Buffalo, N.Y.: Prometheus, 1989).

Chapter 2. Who—or What?—Is in Control Here?

1. To be more precise, by "control" Gibbs means "attempted control," defined as "*overt* behavior by a human in the belief that (1) the behavior increases or decreases the probability of some subsequent conditions and (2) the increase or decrease is desirable. The commission or omission of an act is overt behavior; and a subsequent condition may be the behavior of an organism or the existence, location, composition, color, size, weight, shape, odor, temperature, or texture of some object or substance, be it animate or inanimate, observable or unobservable. However, the condition may or may not take the form of a change . . ." (Jack P. Gibbs, *Control: Sociology's Central Notion* [Urbana: University of Illinois Press, 1989], pp. 23–24; emphasis in original).

2. Paul E. Thompson, "The First Fifty Years—The Exciting Ones," in *A Century of Fisheries in North America*, ed. Norman G. Benson (Bethesda, Md.: American Fisheries Society, 1970), p. 1.

3. Ibid., p. 2.

4. Ibid., p. 5.

5. Ibid., p. 6.

6. Ibid., p. 10.

7. Norman G. Benson, "The American Fisheries Society, 1920–1970," in *Century of Fisheries in North America,* ed. Benson, p. 13.

8. J. L. McHugh, "Trends in Fishery Research," in *Century of Fisheries in North America,* ed. Benson, pp. 25–56.

9. Robert R. Stickney, *Flagship: A History of Fisheries at the University of Washington* (Dubuque, Iowa: Kendall/Hunt, 1989).

10. Quoted in ibid., p. 2.

11. Ibid., pp. 6–7.

12. Ibid., p. 15.

13. For example, in less than five years of living in the region, I have accumulated several boxes of newspaper clippings, magazine articles, and the like about salmon.

14. Bob Mottram, "Salmon Group Wants Nine Species Listed as Endangered," *Olympia News Tribune,* 16 March 1994, p. D4.

15. Fuchs, *Professional Quest for Truth,* p. 26.

16. Robert K. Merton, *The Sociology of Science* (Chicago: University of Chicago Press, 1973).

17. For example, see: Michael E. Soulé, "The Social Siege of Nature," in *Reinventing Nature? Responses to Postmodern Deconstruction,* ed. M. E. Soulé and G. Lease (Washington, D.C.: Island Press, 1995).

18. Netboy, *Columbia River Salmon and Steelhead Trout.*

19. Anthony Giddens, *The Constitution of Society* (Berkeley: University of California Press, 1984), p. 26; see also Alfred Schutz, *The Phenomenology of the Social World* (1932; reprint, Evanston, Ill.: Northwestern University Press, 1960); C. Wright Mills, *The Sociological Imagination* (New York: Oxford University Press, 1959).

20. Michel Callon, "Society in the Making: The Study of Technology as a Tool for Sociological Analysis," in *The Social Construction of Technological Systems,* ed. Wiebe E. Bijker, Thomas P. Hughes, and Trevor Pinch (Cambridge: MIT Press, 1987), p. 83.

21. Mark Granovetter, "The Old and the New Economic Sociology: A History and an Agenda, " in *Beyond the Marketplace: Rethinking Economy and Society,* ed. Roger Friedland and A. F. Robertson (Hawthorne, N.Y.: Aldine de Gruyter, 1990), p. 99.

22. Paul Dimaggio, "Cultural Aspects of Economic Action and Organization," in *Beyond the Marketplace,* ed. Friedland and Robertson, p. 117.

23. National Marine Fisheries Service, 1998, Internet data from: http://remora.ssp.nmfs.gov/agentwebpl/plsql/webst1.MF_ANNUAL_LANDINGS.RESULTS; Department of Fisheries and Oceans, 1998, Internet data from: http://www.ncr.dfo.ca/communic/statistics/landings/S1996pve.htm.

24. Established in 1980 by an act of Congress, the Northwest Power Planning Act created an eight-member Northwest Power Planning Council (NW-

PPC). The NWPPC, comprised of two members each from Washington, Oregon, Idaho, and Montana, acts as a "salmon czar" in the region. Although it was charged with balancing the interests of electrical power interests with those of salmon, until several salmon runs were declared threatened and endangered the NWPPC did little to affect the balance of power, tilted as it was strongly toward electrical generators and users.

25. Netboy, *Columbia River Salmon and Steelhead Trout.*

26. Stickney, *Flagship.*

27. William R. Freudenberg, "The Crude and the Refined: Sociology, Obscurity, Language, and Oil," *Sociological Spectrum,* 17 (1995): 23.

28. Swidler, "Concept of Rationality in the Work of Max Weber," p. 39.

29. Raymond A. Rogers, *Nature and the Crisis of Modernity* (Montreal: Black Rose, 1994).

30. I do not mean to imply that either of these examples, both of which come from the salmon research literature, was the product of bootlegging.

CHAPTER 3. BIOLOGISTS IN THE DRIVER'S SEAT

1. Philip J. Pauly, *Controlling Life: Jacques Loeb and the Engineering Ideal in Biology* (Berkeley: University of California Press, 1987), p. 4.

2. Latour, *Science in Action.*

3. I came to discern *ecosystem* as a label indicative of different *social* systems. In particular, the term ecosystem distinguished those who assert the importance of basic research and who construct salmon as belonging to a "community" of plants, animals, waters—of Nature—as opposed to those with more applied orientations whose exclusive focus was on salmon.

4. John Law, "Technology and Heterogeneous Engineering: The Case of Portuguese Expansion," in *Social Construction of Technological Systems,* ed. Bijker, Hughes, and Pinch, p. 120 (emphasis in original).

5. Latour, *Science in Action,* p. 64 (emphasis in original).

6. See, for example, ibid.; Karin D. Knorr-Cetina and Michael J. Mulkay, eds., *Science Observed: Perspectives on the Social Study of Science* (Beverly Hills, Calif.: Sage, 1983); Bruno Latour and Steve Woolgar, *Laboratory Life: The Construction of Scientific Facts* Princeton: Princeton University Press, 1986).

7. Latour, *Science in Action,* p. 61.

8. Ibid., pp. 238, 240.

9. Scarce, *Eco-Warriors.*

10. *NW Fishletter,* no. 72 (15 December 1998), Internet site at: http://www.newsdata.com/enernet/fishletter/.

11. Eugene P. Odum, "Ecosystem Theory in Relation to Man," in *Ecosystem Structure and Function,* ed. John A. Wiens (Corvallis: Oregon State University Press, 1972).

12. My fieldwork at a salmon biology conference yielded the following two technical definitions of supplementation: (1) "The use of artificial propaga-

tion in an attempt to maintain or increase natural production while maintaining the long-term fitness of the target population and keeping the ecological and genetic impacts on non-target populations within specified biological limits." (2) "The use of eggs, fry, fingerling, and smolt releases of natural stocks to increase production. The intent . . . is to increase the number of recruits per spawner so that, potentially, the harvest rate of the wild stock fisheries can be increased. This could result in greater fishing time and greater fishery stability, an important management objective."

The briefer definition used in the text, and the second of the two technical definitions, may best reflect the dominant, less rigorous approach to supplementation that appears to be commonplace among biologists. I remember one of my early participants laughing at me when I confessed to not knowing what "enhancement" and "supplementation" meant; he did not bother to explain, and when I came to understand the range of definitions for these concepts, I understood his laughter.

13. Charles Perrow, *Normal Accidents* (New York: Basic Books, 1984).

14. "U.S., Canada End Deadlock on Salmon Treaty," *Spokane Spokesman-Review,* 3 July 1994, p. B3.

15. Catton, *Overshoot,* p. 272.

16. "Escapements Hit Record Level as Fisherman By-passed," *Fisherman,* 28 September 1998, Internet site at: http://www.caw.ca/bcsalmon/28sept98.html.

17. Ross Anderson, "Adams River in British Columbia Teems with Salmon," Internet site at: http://www.seattletimes.com/news/local/html98/adam_101898.html.

18. Bruno Latour, "Give Me a Laboratory and I Will Raise the World," in *Science Observed,* ed. Knorr-Cetina and Mulkay, pp. 141–70; also see Zuckerman, "The Sociology of Science," pp. 511–74.

19. Eugene S. Hunn, *Nch'i-Wána, "The Big River": Mid-Columbia River Indians and Their Land* (Seattle: University of Washington Press, 1992).

Chapter 4. Thinking and Making Salmon

1. Arnold Arluke and Clinton R. Sanders, *Regarding Animals* (Philadelphia: Temple University Press, 1996), p. 173.

2. Max Weber, *The Theory of Social and Economic Organization* (New York: Free Press, 1947).

3. J. L. McHugh, "Trends in Fishery Research," in *Century of Fisheries in North America,* ed. Benson, p. 27.

4. John Van Maanen, introduction to *Varieties of Qualitative Research,* ed. John Van Maanen, James M. Dabbs Jr., and Robert R. Faulkner (Beverly Hills, Calif.: Sage, 1982), p. 16; also see Harold Garfinkel, *Studies in Ethnomethodology* (New York: Prentice-Hall, 1967).

5. Van Maanen, introduction.

6. Netboy, *Columbia River Salmon and Steelhead Trout;* see also: Clifford Geertz, *The Interpretation of Cultures* (New York: Basic Books, 1973).

7. Hunn, *Nch'i-Wána, "The Big River."*

8. Arluke and Sanders, *Regarding Animals.*

9. Roderick Nash, *The Rights of Nature* (Madison: University of Wisconsin Press, 1989), p. 18.

10. Marjorie Grene, *Approaches to a Philosophical Biology* (New York: Basic Books, 1968), p. 23.

11. Merchant, ed., *Major Problems in American Environmental History.*

12. Jack P. Gibbs, *A Theory about Control* (Boulder, Colo.: Westview, 1994), p. 39.

13. Don Ihde, *Technology and the Lifeworld: From Garden to Earth* (Bloomington: Indiana University Press, 1990), p. 76.

14. Thompson, "The First Fifty Years," p. 7.

15. As with biologists' names, the names and locations of some hatcheries, rivers, towns, and other geographic features have been changed to protect respondents' confidentiality.

16. Thomas P. Hughes, *American Genesis: A Century of Invention and Technological Enthusiasm, 1870–1970* (New York: Penguin, 1989), p. 3.

17. Melvin Kranzberg, "Technology and History: 'Kranzberg's Laws,' " *Technology and Culture* 27, no. 3 (1986): 548.

18. Lewis F. Carter, Ladd MacAulay, and Catherine M. Coffey, "An Elegant Application of Appropriate Technology: The Sheep Creek Hatchery," *Environmental Management* 10, no. 1 (1986): 53–60.

19. Latour, *Science in Action,* p. 42.

20. Fuchs, *Professional Quest for Truth,* p. 47.

21. "Tribe, Idaho Settle Steelhead Dispute," *Spokane Spokesman-Review,* 31 December 1994, p. B6.

22. *NW Fishletter,* no. 72 (15 December 1998), Internet site at: http://www.newsdata.com/enernet/fishletter/.

23. Geoffrey M. Hodgson, "Why the Problem of Reductionism in Biology Has Implications for Economics," *World Futures* 57 (1993): 76.

24. Catton, *Overshoot.*

25. In his salmon farming handbook, Ian Dore wrote, "In 1989, aquaculture production of chinook salmon worldwide probably exceeded the wild catch worldwide. In 1989, world aquaculture production of all species of salmon reached 225,000 metric tons, not far short of the U.S. average salmon catch over the most recent five-year period of about 297,000 tons. The problem is that the USA has almost no part of this aquaculture production."

This is so, Dore asserts, because commercial seagoing fishers in Alaska have managed to outlaw aquaculture in the state, and "In Washington, Oregon and California, environmental concerns slow or prevent aquaculture development not just of salmon, but of all species" (Ian Dore, *Salmon: The Illustrated Handbook for Commercial Users* [New York: Van Nostrand Reinhold, 1990], pp. 15–16). In Norway, farming of Atlantic salmon is rapidly growing, as it is in

Japan for Pacific salmon. There, after years of treaty fights with the United States and Canada over Japanese fishing in the North Pacific, coho salmon from the United States were imported in 1973, placed in fresh water until 1975, and then placed in seawater pens, "marking the beginning of the cage culture of salmon in Japan." This allows the highly organized and rationalized Japanese commercial fishers, who once plowed the high seas for salmon, to become "fish farmers. They see aquaculture as a way of supplying their market year-round, not just seasonally" (p. 13).

26. Ibid., p. 8.

27. Robert L. Burgner, "Life History of Sockeye Salmon," in *Pacific Salmon Life Histories,* ed. Cornelis Groot and Leo Margolis (Vancouver: University of British Columbia Press, 1991), pp. 3–117 .

28. Ibid.

29. "Outdoor Idaho," 19 April 1992, Moscow, Idaho: KUID.

30. Latour, *Science in Action.*

31. Ibid., p. 90.

32. Courtland L. Smith, *Salmon Fishers of the Columbia* (Corvallis: Oregon State University Press, 1979).

33. Although see R. D. Hume, *A Pygmy Monopolist* (Madison, Wisc.: State Historical Society of Wisconsin, 1961).

34. Lorne T. Kirby, *DNA Fingerprinting: An Introduction* (New York: Stockton, 1990).

35. Weber, *Theory of Social and Economic Organization;* also see David F. Noble, *America by Design: Science, Technology, and the Rise of Corporate Capitalism* (New York: Knopf, 1977).

36. Ihde, *Technology and the Lifeworld,* p. 34.

Chapter 5. Mythology and Biology

1. Riley E. Dunlap, "Limitations of the Social Constructivist Approach to Environmental Problems," paper presented at the Thirteenth World Congress of Sociology, Bielefeld, Germany, 1994.

2. Murphy, *Rationality and Nature,* p. 17.

3. Ibid., p. 20.

4. Stephen Cole, *Making Science: Between Nature and Society* (Cambridge: Harvard University Press, 1992), p. 27.

5. Ibid., p. 30.

6. Thomas S. Kuhn, *The Structure of Scientific Revolutions* (Chicago: University of Chicago Press, 1970).

7. Evernden, *Social Creation of Nature,* p. 31 (emphasis in original).

8. Latour, *Science in Action.*

9. Simmons, *Interpreting Nature,* p. 15.

10. Evernden, *Social Creation of Nature,* p. 23.

11. Roland Barthes, *Mythologies* (New York: Hill and Wang, 1972), pp. 109, 110 (emphasis in original).

12. Evernden, *Social Creation of Nature*, p. 23.

13. Simmons, *Interpreting Nature*, p. 15.

14. Evernden, *Social Creation of Nature*, p. 138.

15. Perrow, *Normal Accidents*.

16. Kenneth Hoover and Todd Donovan, *The Elements of Social Scientific Thinking*, 6th ed. (New York: St. Martin's, 1995).

17. H. M. Collins, "Expert Systems and the Science of Knowledge," in *Social Construction of Technological Systems*, ed. Bijker, Hughes, and Pinch, p. 333.

18. Ibid., p. 345.

19. Perrow, *Normal Accidents*.

20. Scott Sonner, "Fish Experts Say Saving Salmon Would Cost $160M," *Moscow-Pullman Daily News*, 25 January 1995, p. 4A.

21. Perrow, *Normal Accidents*, p. 4.

22. Berger and Luckmann, *Social Construction of Reality;* Schutz, *Phenomenology of the Social World*.

23. Raymond L. Gorden, *Interviewing: Strategy, Techniques, and Tactics*, 4th ed. (Chicago: Dorsey, 1987); Richard G. Mitchell, *Secrecy and Fieldwork* (Newbury Park, Calif.: Sage, 1993).

24. Heinz R. Pagels, *The Cosmic Code* (New York: Simon and Schuster, 1982), pp. 64–65.

25. Ibid., p. 151.

26. Max Weber, *From Max Weber: Essays in Sociology*, trans. and ed. H. H. Gerth and C. Wright Mills (New York: Macmillan, 1958), p. 139.

27. Roderick Nash, *The Rights of Nature* (Madison: University of Wisconsin Press, 1989); Scarce, *Eco-Warriors*.

28. William R. Catton Jr., "Need for a New Paradigm," *Sociological Perspectives* 26 (1983): 3–15; Riley E. Dunlap, "Paradigmatic Change in Social Science: From Human Exemptionalism to an Ecological Paradigm," *American Behavioral Scientist* 24, no. 1 (1980): 5–14.

29. Perrow, *Normal Accidents*.

30. Evernden, *Social Creation of Nature*, p. 56 (emphasis in original).

Chapter 6. Freedom and Self-Determination in Salmon Biology

1. Netboy, *Columbia River Salmon and Steelhead Trout;* Hunn, *Nch'i-Wána, "The Big River."*

2. Orlando Patterson, *Freedom* (New York: Basic Books, 1991), p. xii.

3. Jeffrey C. Alexander, *The Antinomies of Classical Thought: Marx and Durkheim*, vol. 2 of *Theoretical Logic in Sociology* (Berkeley: University of California Press, 1982), p. 11.

4. Quoted in ibid., p. 25 (emphasis in original).

5. Ibid., p. 83.

6. Quoted in ibid.

7. Donald N. Levine, "Rationality and Freedom: Weber and Beyond," *Sociological Inquiry* 51, no. 1 (1981): 5.

8. Alexander, *Classical Attempt at Theoretical Synthesis: Max Weber*, p. 56.

9. Ibid., p. 24.

10. Quoted in ibid., p. 181.

11. Patterson, *Freedom*, p. xiii.

12. A parallel and appropriate phenomenon to note here is the early European construction of the American interior as a "wilderness." In the profoundly tamed and controlled European landscape, *wildness* was exceptional, if extant at all, in the sixteenth and seventeenth centuries. So it was that in the profoundly tamed and controlled European mind that wilderness became a construction meaning *landscapes-that-are-not-controlled* (see Roderick Nash, *Wilderness and the American Mind* [New Haven: Yale University Press, 1982]). That Native Americans had been physically reconstructing vast areas of the American interior for millennia is beside the point; Europeans felt threatened by what they perceived as the uncontrolled other, Nature, around them. Wilderness became the label for the fearsome, foreign, dangerous world that they could not fathom, a world that was civilized to the aboriginal peoples of this continent (see J. Baird Callicott, "The Ethnocentricity of Wilderness Values," in *Major Problems in American Environmental History*, ed. Merchant, pp. 409–12).

13. Here I consciously ignore the widespread control over ecosystems that was practiced by many gatherer-hunter tribes, as for instance by native peoples who used fire to encourage the growth of certain plants. See Johan Goudsblom, *Fire and Civilization* (New York: Penguin, 1992); Carolyn Merchant, *Ecological Revolutions: Nature, Gender, and Science in New England* (Chapel Hill: University of North Carolina Press, 1989).

14. Alston Chase, *Playing God in Yellowstone: The Destruction of America's First National Park* (Boston: Atlantic Monthly Press, 1986).

15. Charles Perrow, *Complex Organizations*, 3d ed. (New York: Random House, 1986), p. 132.

16. Paul K. Feyerabend, *Against Method: Outline of an Anarchistic Theory of Knowledge* (London: NLB, 1975).

17. Stephen R. Fox, *The American Conservation Movement* (Madison: University of Wisconsin Press, 1985); Nash, *Wilderness and the American Mind;* and idem, *The Rights of Nature*.

18. Julian L. Simon and Herman Kahn, eds., *The Resourceful Earth: A Response to "Global 2000"* (New York: Basil Blackwell, 1984).

19. Aldo Leopold, *A Sand County Almanac* (New York: Ballantine, 1966), p. 262.

20. Paul R. Ehrlich, *The Machinery of Nature* (New York: Touchstone, 1986), p. 238.

21. Edward O. Wilson, *The Diversity of Life* (Cambridge, Mass.: Belknap, 1992), p. 228.

22. Reed F. Noss, "Conservation Biology," *Earth First! Journal* 8, no. 1 (1987): 23.

23. Reed F. Noss, "Wither *Conservation Biology*?" *Conservation Biology* 7, no. 2 (1993): 215.

24. Will Nehlsen, Jack E. Williams, and James A. Lichatowich, "Pacific Salmon at the Crossroads: Stocks at Risk from California, Oregon, Idaho, and Washington," *Fisheries* 16, no. 2 (1991): 4.

25. Alexander, *Classical Attempt at Theoretical Synthesis: Max Weber*, p. 99.

26. Mottram, "Salmon Group Wants Nine Species Listed as Endangered," p. D4.

27. Gerald R. Bouck, "Are We Ready for Advocacy?" *Fisheries* 17, no. 4 (1992): 54.

28. Alexander, *Classical Attempt at Theoretical Synthesis: Max Weber*, p. 104.

29. Martin Albrow, *Max Weber's Construction of Social Theory* (New York: St. Martin's, 1990), p. 7.

30. Devall, *Simple in Means, Rich in Ends*, p. 19; see also Devall and Sessions, *Deep Ecology;* and Michael Tobias, ed., *Deep Ecology* (San Marcos, Calif.: Avant, 1988).

Paul Ehrlich argued for an earth-centered religion and specifically embraces deep ecology, writing, "The main hope for changing humanity's present course may lie less with politics . . . than in the development of a world view drawn partly from ecological principles—in the so-called deep ecology movement. . . . The deep ecology movement thinks today's human thought patterns and social organization are inadequate to deal with the population-resource-environment crisis—a view with which I tend to agree. . . . I am convinced that such a quasi-religious movement, one concerned with the need to change the values that now govern much of human activity, is essential to the persistence of our civilization" (Ehrlich, *Machinery of Nature*, p. 17).

It is also worth noting that the founder of the deep ecology perspective, Arne Naess, contributed a chapter in conservation biology cofounder Michael Soulé's book *Conservation Biology* (Sunderland, Mass.: Sinauer Associates, 1986).

31. Murphy, *Rationality and Nature*, p. 243.

32. Peter F. Brussard, Dennis D. Murphy, and C. Richard Tracy, "Cattle and Conservation Biology—Another View," *Conservation Biology* 8, no. 4 (1994): 919, 921.

33. Thompson, "The First Fifty Years," pp. 1–2.

34. Peter Berger, *Invitation to Sociology* (New York: Anchor, 1963), p. 136.

35. Weber, *Protestant Ethic*, p. 181.

36. Ehrlich, *Machinery of Nature*, p. 17.

CHAPTER 7. SALMON WARS AND THE "NATURE" OF POLITICS

1. Matthews, " 'Constructing' Fisheries Management," p. 46.

2. Ibid., p. 47.

3. Ibid., p. 50 (emphasis in original).
4. Quoted in ibid., p. 53.
5. Ibid.
6. Ibid., p. 56.
7. Ibid., p. 57.
8. David Ralph Matthews, *Controlling Common Property: Regulating Canada's East Coast Fishery* (Toronto: University of Toronto Press, 1993), p. 246.
9. Matthews, " 'Constructing' Fisheries Management," p. 51.
10. Canadian Department of Foreign Affairs, 1998, Internet site at: http://www.dfait-maeci.gc.ca/english/geo/usa/salmon.htm.
11. For instance, in my subsequent research on the reintroduction of wolves to Yellowstone National Park, I found that, although journalists make a straightforward ecology/economy divide, some environmental activists actually opposed the wolf reintroduction and some ranchers—who had a great deal of economic interest at stake should the wolves eat their cattle or sheep—favored having wolves literally in their backyards. See Rik Scarce, "What Do Wolves Mean? Conflicting Social Constructions of *Canis lupus* in 'Bordertown,' " *Human Dimensions of Wildlife* 3, no. 3 (1998): 26–45.
12. Pacific Salmon Alliance, "Press Clippings," 1997, Internet site at: http://www.island.net/˜7Ejkurtz/treatypress.htm.
13. Thomas Brandt, "Pacific Salmon Face Uncertain Future," *New York Times*, 28 December 1997.
14. Pacific Salmon Alliance, "Press Clippings."
15. "Fishermen Strip to Protest Salmon Policies," *Bozeman Daily Chronicle*, 15 May 1998, p. 21.
16. Canadian Broadcasting Company, "The Salmon War," 1997, Internet site at: http://newsworld.cbc.ca/news/indepth/salmon.html.
17. Canadian Press, "Clark Explains B.C.'s Stance on Lumber, Salmon to Upset Americans," 8 July 1998, Internet site at: http://interactive.cfra.com/1998/07/08/46176.html.
18. *NW Fishletter*, no. 83 (28 June 1999), Internet site at: http://www.newsdata.com/enernet/fishletter/.
19. Matthews, *Controlling Common Property*, p. 245.
20. Anne Swardson, "Mystery of Vanishing Salmon Puzzles Canadians," *Washington Post*, 31 December 1994, pp. A23ff.
21. Rogers, *Nature and the Crisis of Modernity*, p. 91.
22. Ibid., p. 93.
23. Quoted in Brad Knickerbocker, "Why GOP Seeks to Fillet Salmon Plan," *Christian Science Monitor*, 25 January 1995, p. 2.
24. Quoted in Swardson, "Mystery."
25. Matthews, " 'Constructing' Fisheries Management," p. 57 (emphasis in original).
26. Rogers, *Nature and the Crisis of Modernity*, p. 5.
27. Ibid.

Chapter 8. Constructing Nature—and Experiencing It

1. George Herbert Mead, *Mind, Self and Society* (Chicago: University of Chicago Press, 1934).

2. Quoted in Robert Kuhn McGregor, "Nature over Civilization," in *Major Problems in American Environmental History*, ed. Merchant, p. 203.

3. Annie Dillard, "Living Like Weasels," in *Teaching a Stone to Talk* (New York: Harper and Rowe, 1982), p. 14.

4. Ibid.

5. Evernden, *Social Creation of Nature*, pp. 110, 111 (emphases in original).

6. Barthes, *Mythologies*, p. 110.

Appendix. Methods and Related Literature

1. Groot and Margolis, eds., *Pacific Salmon Life Histories*.

2. Earl Babbie, *The Practice of Social Research*, 7th ed. (Belmont, Calif.: Wadsworth, 1995).

3. Norman K. Denzin and Yvonna S. Lincoln, eds., *Handbook of Qualitative Research* (Thousand Oaks, Calif.: Sage, 1994).

4. Barney G. Glaser and Anselm L. Strauss, *The Discovery of Grounded Theory: Strategies for Qualitative Research* (New York: Aldine de Gruyter, 1967); Kathy Charmaz, "The Grounded Theory Method: An Explication and Interpretation," in *Contemporary Field Research*, ed. Robert M. Emerson (Prospect Heights, Ill.: Waveland Press, 1983), pp. 109–26; Kathy Charmaz, " 'Discovering' Chronic Illness: Using Grounded Theory," *Social Science and Medicine* 30, no. 11 (1990): 1161–72; idem, "Grounded Theory," in *Rethinking Psychology*, ed. Jonathan Smith, Rom Harre, and Luk Van Langenhove (London: Sage, 1995), pp. 27–49; Denzin and Lincoln, eds., *Handbook of Qualitative Research*.

5. Glaser and Strauss, *Discovery of Grounded Theory*, p. 2.

6. John D. McCarthy, foreword to *The Myth of the Madding Crowd*, by Clark McPhail (New York: Aldine de Gruyter, 1991).

7. Patricia A. Adler and Peter Adler, "The Ethnographers' Ball (Revisited)," forthcoming in *Journal of Contemporary Ethnography*.

8. Charmaz, "Grounded Theory," p. 35 (emphasis in original).

9. Bijker, Hughes, and Pinch, eds., *Social Construction of Technological Systems;* Latour, *Science in Action;* Knorr-Cetina and Mulkay, eds., *Science Observed;* Zuckerman, "The Sociology of Science," pp. 511–74.

10. Trevor Pinch and Wiebe Bijker, "The Social Construction of Facts and Artifacts: Or How the Sociology of Science and the Sociology of Technology Might Benefit Each Other," in *Social Construction of Technological Systems*, ed. Bijker, Hughes, and Pinch, p. 27.

11. Malcolm Spector and John I. Kitsuse, *Constructing Social Problems* (Menlo Park, Calif.: Cummings, 1977); John I. Kitsuse and Joseph W. Schneider,

preface to *Images of Issues: Typifying Contemporary Social Problems*, ed. Joel Best (New York: Aldine de Gruyter, 1989).

12. Berger and Luckmann, *Social Construction of Reality*, 20.

13. Pinch and Bijker, "Social Construction of Facts and Artifacts," p. 18.

14. Fuchs, *Professional Quest for Truth*, p. 30.

15. Berger and Luckmann, too, noted the importance of reflexivity to constructivists, although their approach fell short of clarifying the issue. They wrote, "To be sure, the sociology of knowledge, like all empirical disciplines that accumulate evidence concerning the relativity and determination of human thought, leads toward epistemological questions concerning sociology itself as well as any other scientific body of knowledge. . . . How can I be sure, say, of my sociological analysis of American middle-class mores in view of the fact that the categories I use for this analysis are conditioned by historically relative forms of thought . . . ?

"Far be it for us to brush aside such questions. All we would contend here is that these questions are not themselves part of the empirical discipline of sociology. They properly belong to the methodology of the social sciences. . . . The sociology of knowledge, along with the other epistemological troublemakers among the empirical sciences, will 'feed' problems to this methodological inquiry. It cannot resolve these problems within its own proper frame of reference" (Berger and Luckmann, *Social Construction of Reality*, pp. 13–14).

16. The related literature from other disciplines is extensive, though in general it does not speak to the range of concerns raised by sociologists. For example, in anthropology see Mary Douglas, *Purity and Danger: An Analysis of Concepts of Pollution and Taboo* (London: Routledge and Kegan Paul, 1966); idem, *Natural Symbols: Explorations in Cosmology* (London: Barrie and Rockliff, 1970); idem, *Implicit Meanings: Essays in Anthropology* (London: Routledge and Kegan Paul, 1975); idem, *Essays in the Sociology of Perception* (London: Routledge and Kegan Paul, 1982); idem, *Risk Acceptability According to the Social Sciences* (New York: Russell Sage Foundation, 1985); Elizabeth A. Lawrence, *Rodeo: An Anthropologist Looks at the Wild and the Tame* (Knoxville: University of Tennessee Press, 1982); idem, *Hoofbeats and Society: Studies in Human-Horse Interactions* (Bloomington: Indiana University Press, 1985); and idem, *Hunting the Wren: The Transformation of Bird to Symbol: A Study in Human-Animal Relationships* (Knoxville: University of Tennessee Press, 1997). In psychology, see Karl Dake, "Orienting Dispositions in the Perception of Risk: An Analysis of Contemporary Worldviews and Cultural Biases," *Journal of Cross-Cultural Psychology* 22, no. 1 (1991): 61–82; idem, "Myths of Nature: Culture and the Social Construction of Risk," *Journal of Social Issues* 48, no. 4 (1992): 21–37; Karl Dake and Aaron Wildavsky, "Theories of Risk Perception: Who Fears What and Why?" *Daedalus* 119, no. 4 (1990): 41–60. In English, see Philip J. Pauly, *Controlling Life: Jacques Loeb and the Engineering Ideal in Biology* (Berkeley: University of California Press, 1987). In environmental education, see Constance L. Russell, "The Social Construction of Orangutans: An Ecotourist Experience," *Society and Animals* 3, no. 2 (1995): 51–170. In environmental history, see William Cronon, *Changes in the*

Land: Indians, Colonists, and the Ecology of New England (New York: Hill and Wang, 1983); Evernden, *Social Creation of Nature;* Nicholas Jardine, James A. Secord, and Emma C. Spary, *Cultures of Natural History* (Oxford: Oxford University Press, 1996); Merchant, *Ecological Revolutions.* In environmental studies, see Elizabeth Ann R. Bird, "The Social Construction of Nature: Theoretical Approaches to the History of Environmental Problems," *Environmental Review* 11, no. 4 (1987): 255–64; Rogers, *Nature and the Crisis of Modernity.* In geography, see Peter Goin, *Humanature* (Austin: University of Texas Press, 1996); Robert Mugerauer, *Interpreting Environments: Traditions, Deconstruction, Hermeneutics* (Austin: University of Texas Press, 1995); Simmons, *Interpreting Nature.* In philosophy, see Steven Vogel, *Against Nature: The Concept of Nature in Critical Theory* (Albany: State University of New York Press, 1996). Interdisciplinary works include Soulé and Lease, eds., *Reinventing Nature?;* and William Cronon, ed., *Uncommon Ground: Toward Reinventing Nature* (New York: W. W. Norton, 1995).

17. Cronon, *Uncommon Ground.*

18. Dunlap, "Limitations of the Social Constructivist Approach to Environmental Problems"; Murphy, *Rationality and Nature.*

19. Frederick H. Buttel, Ann P. Hawkins, and Alison G. Power, "From Limits to Growth to Global Change: Contrasts (Constraints) and Contradictions in the Evolution of Environmental Science and Ideology," *Global Environmental Change* 1, no. 1 (1990): 57–66; Frederick F. Buttel and Peter J. Taylor, "Environmental Sociology and Global Environmental Change: A Critical Assessment," *Society and Natural Resources* 5, no. 2 (1992): 211–30; John A. Hannigan, *Environmental Sociology: A Social Constructionist Perspective* (New York: Routledge, 1995); Peter J. Taylor and Frederick H. Buttel, "How Do We Know We Have Global Environmental Problems? Science and the Globalization of Environmental Discourse," *Geoforum* 23, no. 3 (1992): 405–16.

20. Martin A. Hajer, *The Politics of Environmental Discourse: Ecological Modernization and the Policy Process* (Oxford: Oxford University Press, 1995).

21. William R. Catton Jr. and Riley E. Dunlap, "A New Environmental Paradigm for Post-Exuberant Sociology," *American Behavioral Scientist* 24, no. 1 (1980): 15.

22. Ibid.

23. Dunlap, "Paradigmatic Change in Social Science," pp. 5–14.

24. William R. Catton Jr., "Need for a New Paradigm," *Sociological Perspectives* 26 (1983): 6.

25. Riley E. Dunlap and William R. Catton Jr., "Struggling with Human Exemptionalism: The Rise, Decline, and Revitalization of Environmental Sociology," paper presented at the annual meeting of the American Sociological Association, Miami, 1993, p. 13.

26. Thomas Greider and Lorraine Garkovich, "Landscapes: The Social Construction of Nature and the Environment," *Rural Sociology* 59, no. 1 (1994): 1.

27. Ibid., p. 8.

28. Ibid., p. 18.

29. Ibid., p. 21.

30. Jan E. Dizard, *Going Wild : Hunting, Animal Rights, and the Contested Meaning of Nature* (Amherst: University of Massachusetts Press, 1994).

31. L.J.H. Hickrod and R. L. Schmitt, "A Naturalistic Study of Interaction and Frame: The Pet as 'Family Member,' " *Urban Life* 11, no. 1 (1982): 55–77; J. E. Nash, "What's in a Face? The Social Character of the English Bulldog," *Qualitative Sociology* 12, no. 4 (1989): 357–70.

32. Gary Alan Fine and Lazaros Christoforides, "Dirty Birds, Filthy Immigrants, and the English Sparrow War: Metaphorical Linkage in Constructing Social Problems," *Social Problems* 14, no. 4 (1991): 375–93; M. Cantwell, "Dogs as Racing Machines," paper presented at the annual meeting of the American Sociological Association, Miami, 1993.

33. Arluke and Sanders, *Regarding Animals.*

34. Gary Alan Fine, "Naturework and the Taming of the Wild: The Problem of 'Overpick' in the Culture of Mushrooms," *Social Problems* 44, no. 1 (1997): 68–88.

35. Arluke and Sanders, *Regarding Animals,* p. 2.

36. Buttel, Hawkins, and Power, "From Limits to Growth to Global Change"; Buttel and Taylor, "Environmental Sociology and Global Environmental Change"; and Taylor and Buttel, "How Do We Know We Have Global Environmental Problems?"

37. Catton and Dunlap, "New Environmental Paradigm"; Catton, "Need for a New Paradigm"; Dunlap, "Paradigmatic Change in Social Science"; see also idem, "Limitations."

38. Loren Lutzenhiser, "Sociology, Energy and Interdisciplinary Environmental Science," *American Sociologist* 25 (1994): 58–79 (emphases in original).

39. Murphy, *Rationality and Nature.*

Index